BRAIN SEIZURE DETECTION AND CLASSIFICATION USING EEG SIGNALS

BRAIN SEIZURE DETECTION AND CLASSIFICATION USING EEG SIGNALS

VARSHA K. HARPALE

VINAYAK K. BAIRAGI

ELSEVIER

ACADEMIC PRESS

An imprint of Elsevier

Academic Press is an imprint of Elsevier
125 London Wall, London EC2Y 5AS, United Kingdom
525 B Street, Suite 1650, San Diego, CA 92101, United States
50 Hampshire Street, 5th Floor, Cambridge, MA 02139, United States
The Boulevard, Langford Lane, Kidlington, Oxford OX5 1GB, United Kingdom

Library of Congress Cataloging-in-Publication Data
A catalog record for this book is available from the Library of Congress

British Library Cataloguing-in-Publication Data
A catalogue record for this book is available from the British Library

ISBN: 978-0-323-91120-7

For information on all Academic Press publications
visit our website at https://www.elsevier.com/books-and-journals

Publisher: Mara Conner
Acquisitions Editor: Chris Katsaropoulos
Editorial Project Manager: Gabriela D. Capille
Production Project Manager: Prasanna Kalyanaraman
Designer: Miles Hitchen

Typeset by VTeX

To My Family and Friends!!!

Contents

Preface

Electroencephalography (EEG) remains the most immediate, simple, and rich source of information for understanding phenomena related to brain electrical activities. The main objective of EEG is to observe abnormalities of brain activity called epileptic seizure. Seizure is a neurological disorder in which too many neurons are excited at the same time, triggered by brain injury or by chemical imbalance. The seizures are predominantly characterized by unpredictable interruptions of normal brain function. A seizure occurs when too many nerve cells in the brain "fire" too quickly, causing an "electrical storm." The EEG signals recorded from epileptic patients are analyzed to monitor behavior and extract signals during seizure onset. The time and frequency domain, wavelet transform, and empirical mode decomposition methods are optimized feature extraction methods and are used in these experiments. One-way analysis of variance (ANOVA) is used as feature selection technique and high-performance machine learning classifiers are applied for classification of EEG signals as normal and epileptic EEG signals.

The book discusses a new feature extraction method, singular spectrum-based empirical wavelet transform (SSEWT), for improved classification of seizures as significant seizure types, specifically epileptic and nonepileptic seizures. The performance of the method is compared with existing feature extraction methods using wavelet transform and empirical wavelet transform. The machine learning classifiers are used for classification of EEG signals as normal, epileptic seizure, and nonepileptic seizure (nonepileptic patients). The contribution of the work lies in the identification of nonepileptic patients using SSEWT. The method is beneficial to obtain high classification accuracies and low false positive rates.

Chapter 1 reviews recent trends in research along with proven methods and identifies their limitations by analyzing the current international status of research in epileptic seizure detection, prediction, and classification. The review motivates the contents of the book.

Chapter 2 explains the methodology used in the book, including EEG signal measurement, the recording setup, and preprocessing. The need of noise and artifact removal using filter methods and independent component analysis (ICA) is discussed and demonstrated with related results.

Chapter 3 covers the seizure detection methods and demonstrates the application and performance of implemented methods. The process includes feature extraction with time domain analysis, frequency domain analysis, wavelet transform, empirical mode decomposition, empirical wavelet transform and the newly proposed method SSEWT, followed by feature selection by ANOVA, and extracted features are classified using a machine learning approach. The chapter covers the implementation of a field-programmable gate array architecture for seizure detection using a genetic algorithm.

Chapter 4 discusses prediction models for the occurrence of seizures using pattern adapted wavelet transform (PAWT) and a fuzzy classifier. The relevant features are extracted, selected, and classified as normal, preseizure, and seizure state of the signal. Thus preseizure detection using PAWT and a fuzzy classifier predicts the occurrence of seizures before seizure onset.

Chapter 5 demonstrates a new efficient method for seizure classification to categorize epileptic and nonepileptic patients. The SSEWT and singular spectrum-based empirical mode decomposition methods are discussed in detail, together with their significance, and a machine learning approach is used to classify the seizure types. The optimized random forest classifier is discussed here with its performance metric.

Chapter 6 discusses the overall results and conclusions observed during seizure detection, prediction, and classification. The scope and limitations of the book are highlighted along with future work.

The results discussed here were obtained using standard labeled scalp and intracranial EEG signals from the MIT CHB Scalp EEG database, collected at the Children's Hospital in Boston, from Goldberger (2013), and from the TUH EEG Seizure Corpus by the American Clinical Neurophysiology Society, USA. The results demonstrated are based on these standard databases and verified by neurologists. For real-time databases, the research was executed with the same specifications for comparable results.

We recommend the book for researchers working in the fields of epileptic seizure detection, prediction, and classification. The contents are also useful to the academician teaching biomedical signal processing.

Varsha K. Harpale
Vinayak K. Bairagi

Acknowledgments

It is with immense gratitude that I acknowledge the support and guidance of my coauthor, **Prof. Dr. Vinayak K. Bairagi**, for encouraging me to write this book on biomedical signal processing. He has always been a source of knowledge and inspiration for me, and the timely completion of my book is thanks to him.

Furthermore, I would like to express my feelings of gratitude towards Pimpri Chinchwad College of Engg. (PCCOE), Pune and all senior faculty members for their constant support and encouragement during the completion of the book. Additionally, special thanks go to AISSMS, Institute of Information Technology, Pune for providing research facilities for experimentations.

I owe an enormous debt of gratitude to all my dear friends who reviewed the book contents multiple times to improve the quality and helped me managing datasets and tools for their constant support. I wish to thank all the neurologists who directly or indirectly validated my work.

Finally, I want to mention my admiration for **my family**, for their constant support, understanding, and stress relieving attitude during the work. **Smt. Anjana Pawar, my mother, Mr. Krishnakant Harpale, my husband, and Mst. Siddhant Harpale, my son**, have made equal contributions during the work and when writing this book, for which I always feel profound gratitude in my heart.

Varsha K. Harpale

CHAPTER 1

Introduction

Engineering research in neurological disorders

1.1 Introduction

Prevalence rates of common neurological disorders including epilepsy, stroke, Parkinson's Disease (PD), and tumors were estimated through a survey conducted across different regions of the world. Considering all variations of 15 neurologic conditions, i.e., Alzheimer disease and related dementia, amyotrophic lateral sclerosis, brain tumors, cerebral palsy, dystonia, epilepsy, Huntington disease, hydrocephalus, multiple sclerosis, muscular dystrophy, PD, spinal cord injury, Tourette syndrome, and traumatic brain injury the surveys are conducted by standard medical agency such as WHO. Fig. 1.1 shows comparative technical research conducted every year for the identification, prevention, and risk analysis of most of the neurological disorders. It summarizes an on-line study published by IEEE Explore and Science Direct for 2017–2018. With respect to the massive burden associated with neurological disorders, it is clear that the neurological services and resources were disproportionately scarce, especially in low-income and developing countries.

The location-based prevalence of neurological disorders is evaluated mainly based on disability-adjusted life in years (DALYs), years of life lost (YLLs), and years lived with disability (YLDs), as stated by the Global Burden of Disease (GBD) study [2]. Statistically, it is observed that 10.2% of global DALYs and 16.8% of global deaths are due to neurological disorders, and between 1990 and 2015, the number of fatalities increased by 36.7%, and DALYs increased by 7.4%. Fig. 1.2 shows the severity of neurological diseases as compared to other disorders affecting human life, and it is observed that 6.3% of total sufferers have a neurological disorder. Thus, 6.3 million people die every year due to neurological disorders [2]. Neurological disorders have a significant impact on disability and mortality worldwide [3]. Epilepsy is also a high-prevalence disease, accounting for 5% of DALYs and 1.3% of deaths. In 2016, 45.9 million patients had epilepsy; the most affected age groups were 5–9 years and >80 years of age. The DALY rate in men was found higher than in women [4]. The World Health Organization, the International League Against Epilepsy, and the International

Brain Seizure Detection and Classification Using EEG Signals
https://doi.org/10.1016/B978-0-32-391120-7.00006-2

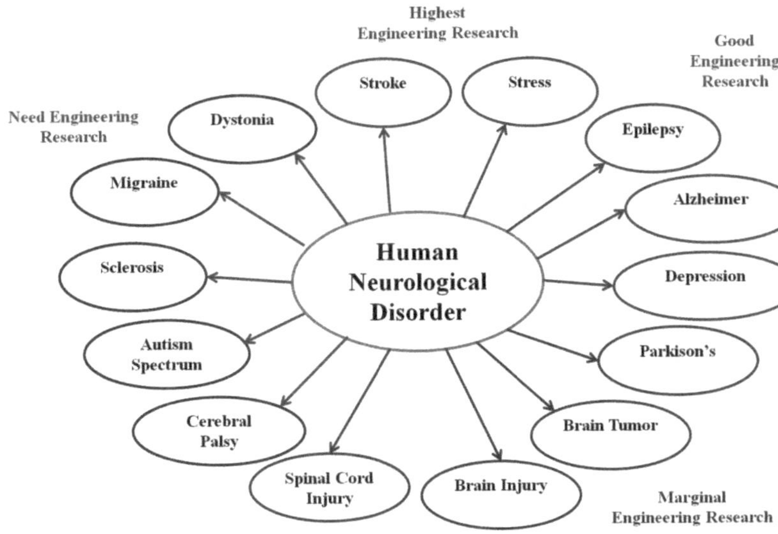

Figure 1.1 Technological research conducted per year in neurological disorders.

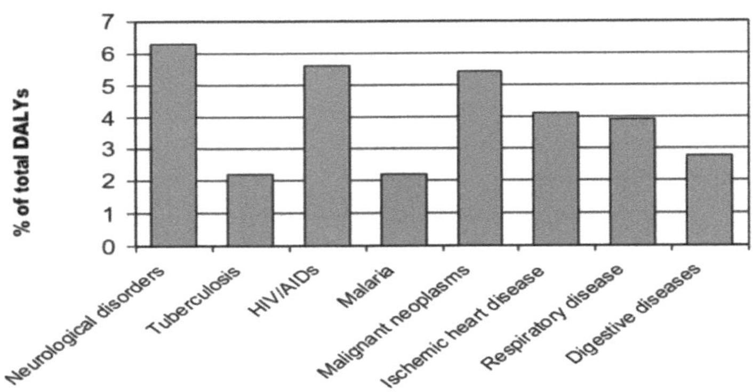

Figure 1.2 Percentage of total DALYs for selected diseases and neurological disorders.

Bureau for Epilepsy operate worldwide, providing technical information about epilepsy and its consequences. They also assist governments and those suffering from epilepsy to reduce the burden of the disorder. Out of the total global disease burden, 1% is due to epilepsy, and 80% of the epilepsy is oberved in the developing world. In some areas, 80% to 90% of people with epilepsy receive no treatment at all [1]. Analysis of data from the Quality and Outcomes Framework suggests a prevalence of 1.15% of diagnosed epilepsy in people aged 18. The prevalence of epilepsy (especially new-onset epilepsy) among the elderly population is increasing, although the possibility of misdiagnosis also remains high [5].

Epilepsy is a chronic noncommunicable disorder of the brain that affects about 40 to 50 million people of all ages in the world [1]. India, with a population of about 1.3 billion, is the largest democracy in the world. Thus to deal with epilepsy in India, awareness of the illness and multiple treatment options are needed desperately. In India, about 10 million people have epilepsy, with a highest prevalence in rural areas and moderate in urban areas. There is a high prevalence of epilepsy among children and young adults as per statistics provided by The Times of India [6].

Epilepsy is a disorder characterized by recurring seizure. Feature extraction, feature analysis, and classification of brain signals are still challenging issues for researchers due to the variations in electroencephalography (EEG) signals. Changes of brain signals depend on the affected areas of the brain, the methods adopted for recording brain signals, the number of channels, and different patterns of EEG signals extracted from different people. Thus, the detection of seizures with high accuracy is still a challenging issue. Prediction of seizures is also a difficult problem. The EEG signal during the preseizure state, the seizure state, and the postseizure state is characterized by preictal, ictal, and postictal periods of the EEG signal. A high predictive capability of seizures (i.e., ictal) and a low (close to zero) false positive rate can be achieved by analyzing interictal and preictal signals. The ictal period starts immediately after the preictal period [7].

Abnormal movements and unconsciousness are the brain states during seizures and are responsible for temporary loss of control. Seizures observed due to sudden abnormal electrical discharges in the brain are referred to as epileptic seizures. Stress-related seizures are called psychogenic nonepileptic seizures (PNES). PNES are similar to epileptic seizures and are hence often misdiagnosed as epileptic seizures. PNES result from the physiological and psychological behavior of the brain. Various nonepileptic seizures (NES) are observed in patients and are usually misdiagnosed and treated with antiepileptic drugs (AEDs). Hence, there is a requirement for the classification of NES and epileptic seizures. PNES look like epileptic seizures but are not caused by abnormal electrical discharges. They are stress-related or "emotional," and result from traumatic psychological experiences, sometimes from the forgotten past. They are sometimes called pseudoseizures, and they are the seizures that are most commonly misdiagnosed as epilepsy [8].

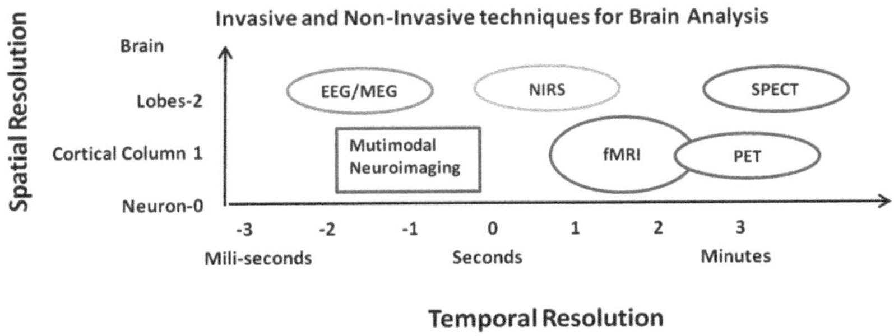

Figure 1.3 Comparison of spatial and temporal resolution of various noninvasive and invasive techniques [10].

1.2 Methods of brain analysis

The brain is a sensitive part of our body that controls various physical activities. It monitors multiple sensations, controls muscular activities, and makes movement, speech, sight, and (facial) expressions possible. There are various methods of brain signal measurement, classified as invasive and noninvasive methods. Invasive methods are associated with a risk of infection due to the penetration of needle-like electrodes in the cortical region of the brain. Usually, invasive techniques are used for surgery or during brain injury monitoring. Noninvasive methods are used for monitoring and recording EEG signals from the scalp. Noninvasive methods are affected by noise and artifacts [9].

Electroencephalography (EEG) is an electrical method for recording brain signals. The EEG signal is recorded from electrodes placed on the scalp or penetrated in the cortex by standard electrode placement methods defined by the International 10-20 system. The EEG is a composite and nonstationary signal which is processed further for diagnosis of disorders. Magnetoencephalography (MEG) is a noninvasive technique that measures the magnetic activity generated by the brain. Both EEG and MEG have an excellent spatial resolution and a low temporal resolution, but MEG is bulky and costly. Fig. 1.3 shows a comparison of various biomedical brain analysis methods used for signal or image processing. Magnetic resonance imaging (MRI) is used to extract information from topographic maps of the structure of the brain. It has a better spatial resolution but a poor temporal resolution, whereas functional MRI (fMRI) has moderate spatial and temporal resolution extracted from metabolic processes in the brain. Near-infrared spectroscopy (NIRS) is a noninvasive method used for medical application. Electromagnetic waves are radiated, and the brain tissues are scanned for absorption in the infrared spectrum, which is indicative of blood hemoglobin and hence neural activity. NIRS is a wireless and portable solution, as opposed to fMRI, and can easily be used for infants as well. fMRI can be used to observe activation of the full brain, whereas NIRS can only be used to scan tissues; hence, fMRI is preferable over NIRS. Positron emission tomography (PET) is a noninvasive imaging technique to examine the brain and is useful to observe brain activity using a radioactive tracer in the bloodstream. The active and nonactive parts of the brain are identified based on the glucose levels utilized by brain cells. It is useful in the diagnosis of diseases such as epilepsy, Alzheimer's disease (AD), PD, and cancer. However, it has a low spatial resolution, it is very time consuming, and it needs a bulky and costly setup; hence, it is not useful in remote places.

Bin He et al. [10] present a comparison of different invasive and noninvasive methods based on the spatial and temporal resolution of each technique. The EEG/MEG methods have a high spatial resolution and a low temporal resolution. EEGs can be analyzed based on the timing and frequency of signals. For brain monitoring at remote places, a real-time, portable, low-weight method is required. For epilepsy monitoring, EEG can be a preferable method, but enhancement of the spatial precision of EEG to achieve high spatiotemporal mapping of the brain signals and to image the coordinated brain networks noninvasively remains a significant challenge. Recent advancements and current challenges in dynamic functional neuroimaging techniques are discussed here, including various methods such as electrophysiological source imaging and multimodal neuroimaging using the integration of fMRI and EEG/MEG. fMRI provides a high spatial resolution in imaging brain activity but has a low temporal resolution.

EEG is a noninvasive technique to measure signals of brain activity and the state of the brain. EEG is used in the monitoring of the brain and in the diagnosis of many neurological illnesses. Current trends in signal processing and machine learning techniques have allowed the automated detection of neurological events for many medical applications. The major challenge observed in the detection and prediction of seizures with EEG signals is to achieve high accuracy for real-time applications. When the detection techniques mentioned above are applied to the EEG signals with different methodologies, their performance is about 80% or less, which is much lower than expected. The advantages of EEG over other methods are:

- EEG is recorded with invasive and noninvasive method, so it is easily selected as per requirements to experiment with human subjects.
- It has a good spatial resolution and time resolution (normally in the millisecond range) which permits to follow the temporal dynamics of brain processes.

The main challenge of EEG is its small signal amplitude in the range of microvolts (μV) and its sensitivity to noise. Various limitations are observed with EEG signals. Firstly environmental conditions and low-frequency electromagnetic fields can cause significant interference to EEG recordings. The second limitation is the activity of muscles in the head region (e.g., eye blinking, chewing, speaking). Other restrictions include a weak signal-to-noise ratio and the need to deal with large subject-specific, inter- and intratrial variability, which demands sophisticated data analysis. To provide solutions to these difficulties, multiple experiments and data analysis methods have been developed. Thus EEG remains the most immediate, simple, and abundant source of information for understanding phenomena related to electrical brain activities [9]. EEG is primarily used to observe abnormalities in brain activity called seizures. A seizure is a neurological disorder in which too many neurons are excited at the same time due to brain injury or a chemical imbalance. Seizures are predominantly characterized by unpredictable interruptions of normal brain function. A seizure occurs when too many nerve cells in the brain fire too quickly, causing an electrical storm. Fig. 1.4 shows EEG signals recorded in the normal state and the seizure state. It is observed that the normal EEG signal varies in the range from 225 to −164 μV, whereas the seizure EEG signal ranges from 1435 to −1283 μV. Epileptic EEG signals can be divided into the following five periods: a nonseizure period, an ictal period, a preictal period, a postictal period, and an interictal period [11]. The EEG signal state is clearly distinguished based on the amplitude and frequency during the preictal, ictal, and postictal periods. Prediction and detection of seizures by analyzing ictal, preictal, and interictal periods can help to identify the subsequent seizure, provide better treatment strategies, and improve safety.

The flow of data analysis using an EEG signal is as follows. Firstly EEG signals are preprocessed for filtering and artifact removal, and then significant features related to healthy and epileptic EEG signals are extracted for seizure detection, seizure prediction, and classification of detected seizures as nonepileptic and epileptic seizures. Further risk analysis is conducted by defining a mathematical model for risk identification.

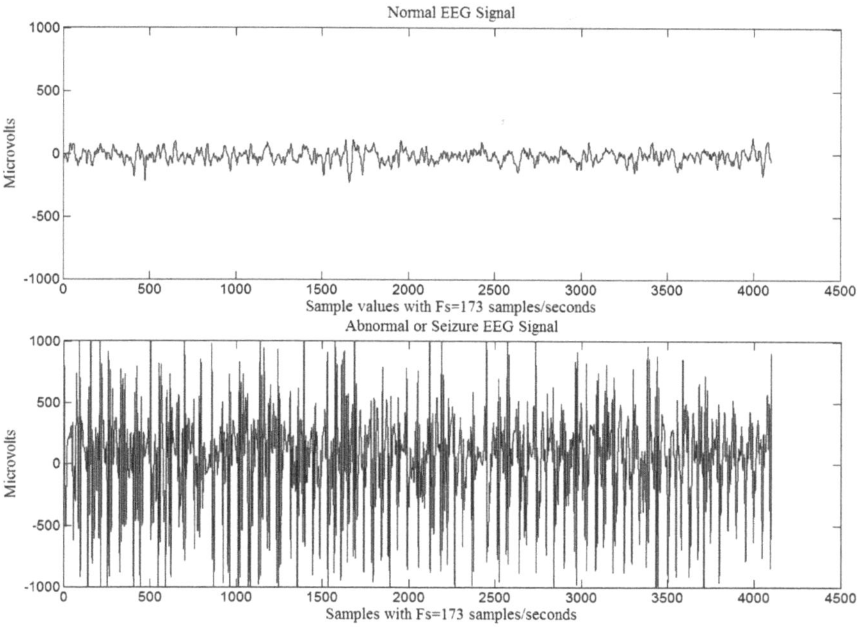

Figure 1.4 Normal and abnormal or seizure EEG signals.

1.3 Recent trends in EEG analysis for feature extraction

The incidence of epilepsy in different regions of India ranges from 967 to 4070 (mean, 2394) per 100,000 inhabitants, which provides a rough estimate of over 30 million people with neurological disorders [11]. Higher prevalences of neurological disorders are found in rural areas, where around 9–10 million people have epilepsy. Globally, approximately 50 million people suffer from epilepsy.

Appropriate feature extraction, selection, and classification of EEG signals are major research problems, which are challenging due to (i) the need to observe brain responses timely and (ii) variation in EEG signals between individuals. The performance of the method depends on the brain locations analyzed, the number of channels used, and differences in signal patterns between different people. Moreover, it is challenging to accurately detect and predict seizures based on EEG signals in real-time. The prediction of seizures is also a difficult problem. A high predictive capability of seizures (i.e., ictal) and a low (close to zero) false positive rate can be achieved by analyzing interictal and preictal signals, assuming that the ictal period starts immediately after the preictal period.

The objectives of the chapters discussed here are (i) to summarize recent results from EEG signal analysis and (ii) to understand (1) the principles underlying the methods adopted for seizure detection, prediction, and classification, for EEG analysis, for EEG signal feature extraction, and for artifact removal, (2) trends in seizure detection and prediction, (3) existing architectures of FPGA models, and (4) methods adopted for (a) classification of epileptic and nonepileptic seizures and (b) risk analysis of detected seizures.

Justin Dauwell et al. [12] comprehensively explain the research areas in brain signal processing. The selection of the right diagnosis method and treatment strategy is always a challenging task. One limitation of brain electrodes is localization, and thus, coverage of the full brain can be one of the research issues. The real-time preprocessing of EEG signals, which includes filtering, adaptive filtering, and blind source separation steps, is an active research area. Feature extraction of EEG signals is followed by a classification method for brain disorder detection.

Omidvarnia et al. [13] introduced a method for statistical analysis of EEG signals recording time-varying information by a generalized partial directed coherence (tv-gPDC) method. In gPDC, orthogonalization of a multivariate autoregressive model was proposed to minimize the effect of respective sources. The distinct features of the system include extraction of nonstationary information, efficient artifact handling, improvement of the signal-to-noise ratio, and its computational efficiency.

Seung-Hyeon Oh et al. [14] presented a feature extraction method based on quantitative analysis using the Hjorth parameter. This method selects the first frequency band, and the Fisher ratio of the Hjorth parameter gives timing parameters of nonstationary EEG signals. Hjorth parameters, activity, mobility, and complexity indicate a static property of the signal in the time domain. The work presented achieved a higher classification accuracy than the short-time Fourier transform (STFT) method.

Mathematical modeling of the time and frequency domain behavior of the EEG signal was discussed, and the need for time and frequency analysis of EEG signals was observed [15,16]. Table 1.1 shows a summary of the recently used time and frequency algorithms for EEG analysis for different applications. It shows time-frequency algorithms have limitations such as reduced time resolution, fixed window size, and moderate accuracy.

S.J. Husain et al. [17] discuss the Hilbert Huang transform method for EEG analysis, which is composed of empirical mode decomposition and

Hilbert spectral analysis. M. Azarbad et al. [18] used Hilbert Huang transforms and singular spectrum analysis in the preprocessing step as an effective noise removal approach and compared it with the well-known wavelet generalized likelihood ratio algorithm.

Carlos Guerrero-Mosquera et al. [19] proposed a new feature extraction approach using EEG signals for the detection of epileptic seizures based on the smoothed pseudo-Wigner–Ville distribution combined with the McAulay–Quatieri sinusoidal model. Features extracted are time duration, frequency components, and energy of the signal. It achieves a high resolution and facilitates the extraction of features.

Sharma et al. [20] provide a study of the use and significance of entropy for epileptic seizure detection. There are various types of entropies applicable to EEG signals, and each has its separate features with respect to the seizure and normal states of the brain. Entropy is a nonlinear parameter and defines the complexity and uncertainty of data. Oliver Faust et al. [21] state the significance of wavelet decomposition for seizure detection. The discrete wavelet transform is more effective for seizure detection and EEG analysis than the continuous wavelet transform. Specifically, the Debeuchies–Morlet wavelet transform is more suitable for EEG signal processing.

Most of the EEG signal processing-related work is presented for brain–computer interfaces and for the detection of diseases such as autism and AD. EEG preprocessing using a filtering method and noise and artifact removal considering the signal as a stationary signal is essential. Wavelet decomposition plays a significant role in separating brain waves. Various other decomposition techniques contribute efficiently to decomposition and feature extraction. Thus the selection of a proper filtering method for noise and artifact removal and an appropriate decomposition method for feature extraction are important steps in brain disease detection using EEG signals.

1.4 Trends in artifact removal and FPGA architecture designs

Epilepsy is considered to be a major neurological disorder. Currently, numerous antiepileptic drugs are available for seizure control, but 30% of epileptic patients either are drug-resistant or suffer from adverse effects [26]. The epileptic seizure is detected, and the seizure is suppressed using an electrical stimulator [27]. High-performance algorithms or devices are required to use electrical stimulation. Earlier research demonstrated various real-time

Table 1.1 Summary of time-frequency algorithms.

Algorithm	Methodology	Experimentation	Application	Advantages	Limitations
Hjorth parameter [14]	Based on selection of principal frequency band and Fisher ratio	BCI Competition 2008 (IV) Graz dataset. STFT and Hjorth parameters used with accuracy improved by 4.4%	BCI applications	Activity and mobility parameters, good time resolution	Good classification accuracy
Rational discrete short-time Fourier transform (DSTFT) [22]	Based on complex sinusoidal basis function without window	EEG database: University of Bonn. Features: mean, median, standard deviation and min/max values of coefficients	Epileptic seizure detection	Simple, fast computational speed, good frequency resolution	Applied to stationary signals, poor time resolution
Fourier transform [23]	Fourier transform applied to consecutive time window	Real-time database of 24-year single patient. ANN used for classification with 91% accuracy	BCI application	Fast computational speed, good localization	Fixed window size, challenge in selection of window size
Discrete wavelet transform with linear classifier [24]	Multiresolution analysis (MRA), continuous and discrete wavelet transform.	EEG database: University of Bonn. Features: mean, median, standard deviation and min/max values of coefficients. Seizure accuracy 99.5%	Epileptic seizure detection	Flexible window size, better time and frequency resolution	Selection of mother wavelet function (MWF), same MWF for all frequencies

continued on next page

Table 1.1 *(continued)*

Algorithm	Methodology	Experimentation	Application	Advantages	Limitations
Gabor wavelet transform [25]	Modified Gabor wavelet transform	Gabor-based features and PNN classifier with accuracy of 64.78%	Emotion recognition	Adequate time-frequency resolution for sleep EEG	Not specified
Hilbert–Huang transform [17]	Combination of empirical mode decomposition and Hilbert spectral analysis	EEG database: University of Bonn. Accuracy 98.5%	Epileptic seizure detection	Fast diagnosis, high accuracy, good sensitivity and specificity	Mostly applicable to narrow band signals
Smooth Wigner–Ville distribution (SWVD) [19]	Based on McAulay–Quatieri sinusoidal model	EEG database: Clinica Universitaria de Navarra, Department of Neurophysiology. Average sensitivity: 73.2%, average specificity: 95.8%	Epileptic seizure detection	High resolution	Influenced by cross-term interference

seizure detection methods [28], but further improvement is expected. Since artifacts such as owing to eye blinking and muscle movements affect the accuracy of epileptic seizure detection, it is necessary to separate them from EEG signals of interest. The principle of artifact removal using independent component analysis (ICA) [29] is based on the decomposition of mixed signals recorded from different sources. ICA processor implementation with a closed-loop system or a genetic algorithm architecture supporting optimized feature extraction can be designed to enhance the performance of epileptic seizure detection [30,31].

Chen et al. [32] presented a system-on-chip (SoC) architecture, designed using ICA and a lossless data compression method for signal analysis. This architecture is complexity-efficient with an average correlation of 0.9044 and power consumption optimized by 41.6%. Jenihhin et al. [33] describe an original patented algorithm for SASI calculation. For the proof of concept, it was first implemented as a MATLAB® program running on a computer connected to a commercial EEG signal capturing equipment.

W.M. Chen et al. [34] put forth a solution for epilepsy seizure detection. The proposed SoC is a combination of the signal-processing closed-loop system with an eight-channel intracranial EEG (iEEG) and a 10-bit analog-to-digital converter (ADC). It also includes an electronic stimulator based on an adaptive high voltage and a wireless transmission link. The improved epileptic seizure detection method proposed by Yang et al. [35] uses ICA for multichannel signals.

EEG signal analyzers provide effective detection of the mental disorders of a human being. This process became more successful, faster, and more accurate through training and practice. The EEG signal analyzer is used for the analysis of normal and abnormal EEGs. Similarly, EEG analyzers are used to characterize the brain activities associated with particular diseases like depression, seizure, etc. The efficiency and accuracy of real-time systems can be improved by designing biosignal processors using field-programmable gate arrays (FPGAs) for analysis of onset EEG data. Table 1.2 reviews the significance of ICA for artifact removal. It shows that ICA is most commonly used for artifact removal and used in BCI and epileptic seizure detection. Various ICA-based hardware architectures and mathematical models are also useful in different signal processing techniques. The design flow of FPGA implementation is as follows:

- Select specification with consideration of the implementation using codesign with MATLAB and FPGA.
- Simulate design at the behavioral level in VHDL.

Table 1.2 Application of ICA for artifact removal from EEG signals.

Authors	Methodology	Application and performance parameters
Jui-Chieh Liao et al. [36]	System-on-chip architectural design with 90 nm CMOS technology for online recursive ICA (ORICA)	Power consumption, 8.56 mW; latency, 0.2532 sec
M.H. Soomro et al. [37]	Empirical mode decomposition and canonical correlation analysis	2.2 dB to 6.0 dB improvement in signal-to-artifact ratio (SAR)
S. Jirayucharoensak et al. [38]	Neurofeedback system for artifact removal using ICA and wavelet transform	Attention analysis with good artifact removal
J. Liao et al. [39]	ORICA for artifact removal	BCI application
C. Bedoya et al. [40]	Fuzzy classification for multivariate data analysis along with ICA	BCI application
A. Zachariah et al. [41]	Artifact removal with hybrid model of wavelet transform and ICA	Increase in accuracy and reduction in processing time achieved
N. Mammone et al. [42]	Hybrid model of ICA and Hurst exponent	Epileptic seizure detection with 87.9% detection rate
U. Heute et al. [43]	Combination of ICA and adaptive Wiener filter is used	Improved artifact rejection ratio
R. Mahajan et al. [44]	Combination of ICA, modified multiscale entropy (mMSE), and wavelet denoising	Average sensitivity, 90%; average specificity, 98%; average execution time, 0.06 sec

- Synthesizable design at the register-transfer level (RTL) in VHDL.
- Final implementation on FPGA.

Wei-Ming Chen et al. [45] presented a solution for epilepsy seizure detection. The proposed system has an eight-channel closed-loop neural-prosthetic SoC composed of a closed-loop signal-processing path. An eight-channel iEEG acquisition unit consisting of analog amplifiers and a 10-bit delta-modulated ADC, which is used for preprocessing, and these signals are further processed with a programmable biosignal processor including an

Table 1.3 EEG analysis methodologies.

Authors	Methodology	Applications
C. Chen et al. [32]	Complexity-efficient ICA architecture with compression	Brain monitoring system
M. Jenihhin et al. [33]	Real-time EEG analyzer prototype	Brain disorder diagnosis
Wei-Ming Chen et al. [45]	Energy-efficient 10-b delta-modulated SAR ADC and a configurable biosignal processor	Real-time epileptic seizure control
W. Fang et al. [46]	ICA-based SoC architecture of biomedical processors	Tomography system for brain imaging

adaptive high-voltage-tolerant stimulator and a wireless transmission link including a Med Radio-band transceiver along with a wireless power supply system.

W. Fang et al. [46] proposed a low-power biomedical SoC architecture for a portable brain and heart monitoring system. It includes fMRI, EEG, and ECG signal analysis with a tomography system and ICA.

Table 1.3 shows some of the EEG signal analysis methodologies adopted by various researchers. These architectures give a basic idea of the implementation of various EEG signal analyzers. Kuo-Kai Shyu et al. [47] proposed an FPGA-based FastICA architecture to separate independent features from the composite EEG signal and a floating-point architecture to improve accuracy and dynamic performance. A fixed-point FPGA-based architecture was proposed by W. Haung et al. [48], which includes four-channel on-line EEG signal separation and a wireless module implemented with a Bluetooth protocol. The work achieves flexible communication with high speed at 68 MHz sampling rate.

T.J. Chen et al. [49,50] proposed a low-power SoC architecture for epileptic signal processing, analysis, and seizure detection. It is composed of an analog front end module, programmable gain amplifiers, 16-bit sigma-delta-modulated ADC, a 32-bit RISC processor, and a data handling unit. This architecture is energy-efficient and achieves a low power of 28.6 μW at 25 MHz.

M. Shoaib et al. [51] introduced the concept of compressed sensing in the design of a seizure detection processor for real-time EEG analysis. Its performance is compared with that of a local computation method and improvements in detection performance such as detection accuracy, sensitivity, and specificity were found. An SoC architecture for a seizure classification processor with eight-channel EEG acquisition is presented. It is constructed with analog front end module and a distributed Quad-LUT filter architecture for improving classification accuracy and minimizing area and the power delay product. Support vector machine (SVM) classifiers can be well suited for the classification of seizures. Therefore, the architecture uses the SVM algorithm for application design.

1.5 Seizure detection techniques

The field of seizure identification is refined to identify either focal or generalized seizures accurately. Various methods and algorithms were adopted in this field to enhance performance parameters, such as detection or classification accuracy, sensitivity, and specificity.

Kazutaka Jin et al. [52] review various methods used for automatic seizure detection and specify a strong need for a system to enable speedy and accurate reading. One of the observed limitations is that EEG interpretation for seizure detection is very time consuming. The inclusion of lower frequencies (2.5–3 Hz) in the analysis range and additional temporal basal electrodes in the seizure detection montage are critical to improving the detection rate.

Kai-Fu et al. [53] presented a new method for seizure classification using Hilbert–Huang transforms and SVM. This method uses features of the theta, alpha, beta, and gamma waves with RBF-SVM, which provide a higher classification accuracy in the classification of seizure and non-seizure EEG signals. Sleep and related heart diseases can be classified by this method. Limitations of this method include time and computational complexity. Parvez, Paul et al. [55] presented a generic seizure detection approach for feature extraction of ictal and interictal signals using various established transformations and decomposition methods using high-frequency components. This method achieved a significant improvement in average sensitivity of up to 91.36%. Zainuddin Z, La KH, Ong et al. [56] presented a wavelet neural network (WNN)-based system that operates in four stages: feature extraction, feature selection, classification using WNN, and performance analysis. They proposed a novel and reliable seizure detection, where the statistical features are extracted from the discrete wavelet transform in conjunction with the wavelet neural network.

Freeman, W.J. et al. [57] proposed a combination of EEG and MEG to improve analysis and recording by spatiotemporal patterns of neural activ-

Table 1.4 Summary of literature in the field of seizure detection.

Authors	Presented work	Methodology	Parameters observed
Kiranyaz et al. [54]	Classification of patient-specific seizures	Particle swarm optimization (PSO), conditional mutual information maximization (CMIM)	Average sensitivity, 89%; Average specificity, 93%
Chen L.L. et al. [58]	Epileptic seizure detection	Wavelet-based nonlinear features and extreme learning machine (ELM)	Sensitivity, 92%; false detection rate (FDR), 78%
Pachori R. et al. [59]	Epileptic seizure classification	Empirical mode decomposition, artificial neural network classifier	Accuracy, 97.75%
Lee et al. [60]	Classification of normal and epileptic seizure	Wavelet transform and Euclidean distance	Accuracy, 98%; sensitivity, 96%
Yatindra Kumar et al. [61]	Epileptic seizure detection	DWT, entropy, and SVM	Accuracy, 98%
Varun Joshi et al. [62]	Epileptic seizure classification	Fractional linear prediction (FLP), SVM	Classification accuracy, 95.33%
Xu X. et al. [63]	Classification of focal epilepsy and nonepileptic seizure	Fisher discriminate analysis (FDA) and SVM classifier	Classification accuracy, 92%; specificity, 80%

ity. Localizing neuronal activity in the brain, both in time and in space, is a central challenge to improve our understanding of brain function. Localizing neural activity from EEG or MEG data is called electromagnetic source imaging (EMSI). EEG and MEG each provide data with high temporal resolution (measured in milliseconds), but with limited spatial resolution.

Various recent trends and techniques for seizure detection with performance parameters are summarized in Table 1.4. Due to variations in EEG signals, following limitations of real-time systems are observed, such as moderate specificity, long detection times, high computational complexity, and low classification accuracy. Also, the need for a real-time system for seizure detection is observed to handle seizures at remote places.

1.6 Classification of epileptic and nonepileptic seizures

An NES is a period of unusual brain behavior that resembles seizure behavior but is not an epileptic seizure. NES are observed due to low blood pressure or a chemical imbalance, but normally, they are due to psychological problems [64]. NES occur because of problems with handling thoughts, memory, emotions, or sensations in the brain. Such problems are sometimes related to stress [65,66].

NES features such as frequency, triggers, and duration vary from epileptic features. The term "NES" covers both physiologically and psychologically mediated paroxysmal events similar to epileptic manifestations. The psychogenic term is added to NES to indicate psychologically mediated nonepileptic events [67].

Abnormal movement and unconsciousness are responsible for the generation of seizures, which in turns cause temporary loss of control. Seizures observed due to sudden abnormal electrical discharges in the brain are epileptic seizures. Stress-related seizures are referred to as PNES. PNES are similar to epileptic seizures and are hence most often misdiagnosed as epilepsy. PNES result from physiological and psychological behavior of the brain. In the present chapter, we review recent trends in the diagnosis of PNES and classification techniques used to classify epileptic seizures and PNES.

Pippa, E. et al. [68] reported work related to the classification of epileptic and nonepileptic events from multichannel EEG data. Time and frequency domain features are extracted and used for classification. Nonepileptic events are classified as PNES and vasovagal syncope (VVS). Bayesian network, random committee, and random forest classifiers were used with 11 subjects and evaluated, and accuracies of 86%, 83%, and 74% were obtained, respectively. Time and frequency domain features were extracted and used for classification using the WEKA tool.

K. Turner et al. [69] proposed a method which uses mean and standard deviation features to analyze and classify epilepsy and PNES in 66 patients. This study improved our understanding of the impact of PNES and helped providing appropriate clinical, psychological, and social care.

J.A. Pillai et al. [70] elaborated on epileptic and nonepileptic seizures and provided information about the selectivity of electrode location while undergoing video-EEG. This study also discusses interictal EEG characteristics in a patient with epilepsy and PNES such as frontal, temporal, parietal, central, and occipital epileptiform. It suggests that frontal lobe localization could be used to improve the performance in the future.

The review article by J. Durrant et al. [71] discusses diagnosis and prediction of PNES and probable factors required for the diagnosis of PNES. In adults, the scenario is unfortunate with a reduction rate of seizures of around 40%, but in children, it is better. The research study conducted shows that video-EEG or ambulatory EEG can often be diagnosed as PNES.

P. Xu et al. [63], developed a resting scalp EEG-based approach for identifying PNES and epileptic seizures using the common spatial pattern recorded from the brain. An accuracy, sensitivity, and specificity of 92%, 100%, and 80% for classification are reported. The newly proposed spatial pattern network of resting EEG can provide sufficient data to classify seizures as PNES or epilepsy. Fisher discriminate analysis and SVM are used as trained classifiers. EEG recorded here uses the 10–10 international standard system for the placement of electrodes, with a sampling rate of 512 Hz.

Gubbi J. et al. [72] reported the development of a portable device for diagnosis of convulsive PNES with a very high seizure detection rate (100% sensitivity) and a false positive rate of only 6.67% based on 27 patients. The methodology included here uses discrete wavelet transform with mother wavelets db3, db5, coif3, and sym3 and is optimal for the classification by SVM or K-means clustering.

Bayly J. et al. [73] made use of limb movement recorded using an accelerometer and its time-frequency mapping to differentiate between convulsive epileptic seizures and PNES. In this paper, 35 subjects were tested using video-EEG monitoring. It was found that 26 subjects suffered from PNES, 8 suffered from epileptic seizures, and in one subject, both types of seizures were observed. Coefficient of variation (CoV) features were useful in PNES to observe limb movement artifacts. Classification accuracy in this work was 91%, with improved sensitivity and specificity. Kotsopoulos I. et al. [74] review comparative studies of the incidence of epilepsy and predictive factors of epileptic and nonepileptic seizures.

In work [75], the author compares ictal heart rate variability (HRV) parameters as a reflection of autonomic nervous system (ANS) tone in epileptic seizures and PNES and explores differences between interictal and ictal ANS tone in both patient groups. Statistical analysis used in this work includes the mean, standard deviation, and logistic regression analysis for feature extraction and the chi-square and *t*-tests to assess clinical features for distinction between epileptic and nonepileptic seizures. The methodology was tested with 42 patients, and PNES was detected in 24 patients.

1.7 Seizure prediction techniques

Journey et al. [76] reported the need for early detection of seizures and accurate methods adopted by prediction models. Microelectrodes, high-frequency oscillations, a high sampling rate, high-density arrays, and modern analysis techniques are all elements of the recording and detection process that in combination with modeling studies can provide new insights into the dynamics of seizure onsets. This will help to implement improved detection devices that will favorably impact the quality of life of patients.

S. Ramgopal et al. [77] review various techniques and studies of seizure detection and prediction. Various methods are adopted and tested for performance measures such as sensitivity, accuracy, specificity, and the false positive rate. Challenges in automatic detection and prediction systems and closed-loop warning systems remain.

Ali Yadollahpour et al. [78] review recent techniques and challenges in seizure prediction, including various frequency-based methods, nonlinear analysis, statistical analysis, and intelligent systems. It is stated that seizures can be predicted well before onset, facilitating prevention before any clinical medication.

Table 1.5 summarizes research in seizure prediction. Clearly, there is a need for a generalized algorithm that is database-independent and applicable to all types of data, exhibits good performance, and is capable of automated real-time solution for prediction.

1.8 Risk analysis of seizures

L.J. Bonnett et al. [89] provided a clinical study of patient risk based on age and occurrence of seizures. If the patient is seizure-free for more than 6 months, then the risk of death is 20%. Often this is due to the application of a suitable antiepileptic drug.

R. Sukanesh et al. [90] proposed a technology for epilepsy risk level classification using a fuzzy aggression method. Parameters include energy, variance, peaks, events, duration, and covariance. Risk is classified using a fuzzy aggression method as low, medium, high, or very high. Based

Table 1.5 Summary of literature on seizure prediction.

Authors	Database	Methodology	Parameters
N. Ozdemir et al. [79]	Intracranial: Freiburg EEG database	Hilbert–Huang transform (HHT) and Bayesian classifiers	Sensitivity: 96.55%; false positive rate: 21%
A. Petrosian et al. [80]	Intracranial and scalp EEG with 200 Hz sampling rate	Recurrent neural networks (RNNs) and wavelet Decomposition	30–90 sec preictal time
B. Direito et al. [81]	European Database	mRmR and non-dominated sorting genetic algorithm (NSGA-II)	Sensitivity: 99.87%; specificity: 62.30%; accuracy: 66.87%
C.Y. Chiang et al. [82]	Freiburg, CHB-MIT EEG Database and National Taiwan University Hospital	Bivariate patterns method, on-line retraining method	74.2% sensitivity on the ECoG database and 52.2% on the scalp EEG database; false positive rate, 0.2/hr.
K.A. Helini Kulasuriya et al. [83]	Freiburg Database	Wavelet transform and ANN classifier	Sensitivity: 85.7%; specificity: 87.5%; accuracy: 86.67%
James R. Williamson et al. [84]	Freiburg EEG dataset	Novel multivariate EEG features used, SVM	Sensitivity: from 85% to 95%; false positive rate: 18%
C. Teixeira et al. [85]	European EPILEPSIAE project	SVM, multilayer perceptron (MLP) neural networks, and radial basis function (RBF) neural networks	Sensibility: 33% to 100%; false positive rate: 10% to 15%
S. Consul et al. [86]	CHB-MIT Scalp EEG Database	Phase-locking statistics (PLS)	Sensitivity: 88.2%
A. Sharma et al. [87]	CHB-MIT Scalp EEG Database	Synchrony analysis: coherence and correlation of EEG signals	Preictal time, 1–5 sec before arrival
M.Z. Parvez et al. [88]	Freiburg EEG data set	Spatiotemporal relationship of EEG signals using phase correlation	Prediction accuracy: 91.95%; false positive rate: 21.4%

on selected features such as duration, variation in duration, and energy, this method achieves performance index and quality values of 96.56% and 23.02%.

R. Harikumar et al. [91] presented a risk level classification of epileptic patients using a hybrid model of a fuzzy decision tree and a neural network. A quality value of 22.43% was achieved. Epileptic risk in a diabetic patient was also analyzed using fuzzy classification.

J.A. Craner [92] provides a clinical review of seizure severity analyses. A quantitative assessment based on seizure components was performed. Parameters considered here are seizure frequency, seizure type, seizure duration, postictal events, postictal duration, seizure cluster, and physical observations during a seizure. Based on these parameters, scores are calculated, severity is rated, and treatment is suggested.

F. Yaghouby et al. [93] present a seizure risk prediction method based on EEG signals. The subjects are dogs, and the recording method is intracranial. The parameters used are duration severity and false positive rate. S. Gurumurthy [94] classified the brain signals considering epilepsy risk using SVM. The system performance is evaluated with a performance index (PI) rather than a quality value.

1.9 Summary of seizure detection, prediction, and classification

The present chapter can be summarized as follows:

a. Epileptic seizure detection and prediction still poses challenges. The evaluation of most detection and classification techniques is based on accuracy, sensitivity, and specificity.

b. The best achieved values of performance measures are summarized in Table 1.6.

c. Classification of epileptic and nonepileptic seizures is still a matter of concern.

d. Misdiagnosing NES leads to the unnecessary use of antiepileptic medication and affects the patient's learning or working ability.

e. Prediction of seizures long before seizure onset can help patients to take care of them, and thus a real-time prediction model is needed to improve patient safety.

f. A feasible method to detect seizures, estimate seizure severity, and analyze patient risk at remote places is also urgently needed.

g. A portable seizure prediction device is in high demand.

Table 1.6 Summary of the overall literature survey.

Research work	Performance parameters	Highest achieved values	Observations
Seizure detection	Sensitivity	92%	Specificity low.
Seizure prediction	Sensitivity	96%	FPR high, limited to database.
	false prediction rate	0.1/hour	Varies with number of seizures.
	Preictal time	30–90 sec	Prediction time can be improved.
Seizure classification	Accuracy of normal and epileptic seizures	98%	Classification of epileptic and nonepileptic not considered.

The present work focuses on providing solutions by designing and implementing efficient epileptic seizure prediction, identification of NES, and risk analysis of detected seizures.

The objectives of the chapters included in the book are:

1. to study the various methods of time-frequency analysis of brain signals such as wavelet transform and empirical mode decomposition and compare them based on their performance measures [15,16];
2. to study and implement an artifact removal method to improve the performance of the system [95];
3. to propose a system to timely predict seizures and implement an efficient algorithm for seizure detection with high classification accuracy and sensitivity and the capability to localize seizures [96];
4. to analyze detected seizures for severity and risk;
5. to propose and implement a detection model with ASIC prototype; and
6. to propose an improved classification model for epileptic and nonepileptic seizure (or PNES) classification and analyze them based on computational complexity and time complexity.

References

[1] World Health Organization (WHO). Media Centre, Epilepsy, Fact Sheet, http://www.who.int/mediacentre/factsheets/fs999/en/ (accessed 2013).

[2] WHO 2019, Epilepsy factsheet, https://www.who.int/news-room/fact-sheets/detail/epilepsy, 2019, pp. 1–5.

[3] V.L. Feigin, A.A. Abajobir, K.H. Abate, F. Abd-Allah, A.M. Abdulle, S.F. Abera, I. Aichour, Global, regional, and national burden of neurological disorders during

1990–2015: a systematic analysis for the Global Burden of Disease Study 2015, The Lancet Neurology 16 (11) (2017) 877–897.

[4] E. Beghi, G. Giussani, F. Abd-Allah, J. Abdela, A. Abdelalim, H.N. Abraha, A. Awasthi, Global, regional, and national burden of epilepsy, 1990–2016: a systematic analysis for the Global Burden of Disease Study 2016, The Lancet Neurology (2019) 357–377.

[5] National clinical guideline centre: epilepsies methods, evidence and recommendations, Partial Pharmacological Update of Clinical Guideline 20 (2012) 21–28, NICE, www.nice.org.uk.

[6] Japbani Nanda, An international challenge: epilepsy in India, in association with the times of India, N.p., 12 Sept. 2012, Web. 15 Sept. 2013.

[7] M.Z. Parvez, M. Paul, Classification of Ictal and Interictal EEG signals, in: IASTED Conference on Biomedical Engineering, 2013, pp. 791–831.

[8] S.R. Benbadis, L. Heriaud, Psychogenic (non-epileptic) seizures, a guide for families and patients, pp. 3–5, http://hsc.usf.edu/COM/epilepsy/PNESbrochure.pdf, 2012.

[9] S. Snyder, X.A. Shen, A review of brain signal processing methods, in: Int'l Conf. Biomed. Eng. Sci., vol. BIOENG'17, 2017, pp. 10–16.

[10] Bin He, Lin Yang, Christopher Wilkeand, Han Yuan, Electrophysiological imaging of brain activity and connectivity – challenges and opportunities, IEEE Transactions on Bio-Medical Engineering 58 (7) (2011) 1918–1931.

[11] M. Gourie-Devi, Neurological practice: an Indian perspective, Annals of Indian Academy of Neurology 9 (2) (2006) 129–130.

[12] Justin Dauwels, Francois Vialatte, Topics in brain signal processing, in: Asia-Pacific Signal and Information Processing Association Annual Summit and Conference (APSIPA ASC 2010), 2010, pp. 31–36.

[13] A. Omidvarnia, G. Azemi, B. Boashash, John M. O'Toole, Paul B. Colditz, S. Vanhatalo, Measuring time-varying information flowing scalp EEG signals: orthogonalized partial directed coherence, IEEE Transactions On Biomedical Engineering 61 (3) (2014) 680–893.

[14] S.H. Oh, Yu Ri Lee, H.N. Kim, A novel EEG feature extraction method using hjorth parameter, International Journal of Electronics and Electrical Engineering 2 (2) (2014) 106–110.

[15] V.K. Harpale, V.K. Bairagi, Time and frequency domain analysis of EEG signals for seizure detection, in: Proceeding of the IEEE International Conference on Microelectronics, Communication & Computing (MicroCom 2016), NIT, Durgapur, 2016.

[16] V.K. Harpale, V.K. Bairagi, Mathematical modeling and data analysis of electroencephalographic signals: a review, International Journal of Emerging Technology and Advanced Engineering 5 (3) (2015) 352–356.

[17] S.J. Husain, Dr.K.S. Rao, An artificial neural network model for classification of epileptic seizures using Huang-Hilbert transform, International Journal on Soft Computing (IJSC) 5 (3) (2014) 23–33.

[18] M. Azarbad, H. Azami, S. Sanei, A. Ebrahimzadeh, A time-frequency approach for EEG signal segmentation, Journal of AI and Data Mining 2 (1) (2014) 63–71.

[19] C.G. Mosquera, A.N. Vazquez, New approach in features extraction for EEG signal detection, in: Annual International Conference of the IEEE EMBS Minneapolis, 2009, pp. 13–16.

[20] R. Sharma, R.B. Pachori, U.R. Acharya, Application of entropy measures on intrinsic mode functions for the automated identification of focal electroencephalogram signals, Entropy 17 (2015) 669–691.

[21] O. Faust, U.R. Acharya, H. Adeli, A. Adeli, Wavelet-based EEG processing for computer-aided seizure detection and epilepsy diagnosis, Seizure 26 (2015) 56–64.

[22] K. Samiee, P. Kovacs, M. Gabbouj, Epileptic seizure classification of EEG time-series using rational discrete short time Fourier transform, IEEE Transactions On Biomedical Engineering 62 (2) (2014) 541–552.

[23] Abdul-Bary Rouf Suleiman, Toka Abdul-Hameed Fatehi, Feature Extraction Techniques of EEG Signal for BCI Applications, Faculty of Computer and Information, Engineering Department, College of Electronics Engineering, University of Mosul, Iraq, 2007, pp. 35–47.

[24] Thasneem Fathima, M. Bedeeuzzaman, Omar Farooq, Yusuf U. Khan, Wavelet based features for epileptic seizure detection, MES Journal of Technology and Management 2 (1) (2011) 108–112.

[25] S. Nasehi, H. Pourghassem, An optimal EEG-based emotion recognition algorithm using Gabor, WSEAS Transactions on Signal Processing 3 (8) (2012) 87–99.

[26] P. Kwan, S.C. Schachter, M.J. Brodie, Current concepts: drug resistant epilepsy, The New England Journal of Medicine 365 (10) (2011) 919–926.

[27] C.Y. Lin, W.L. Chen, M.D. Ker, Implantable stimulator for epileptic seizure suppression with loading impedance adaptability, IEEE Transactions on Biomedical Circuits and Systems 7 (2) (2013) 196–203.

[28] W.M. Chen, A fully integrated 8-channel closed-loop neural prosthetic SoC for real-time epileptic seizure control, in: Int. Solid State Circuits Conf., 2013, pp. 286–287.

[29] A. Hyvärinen, J. Karhunen, E. Oja, Independent Component Analysis, Wiley, 2001, pp. 1–31.

[30] C.H. Yang, Y.H. Shih, H. Chiueh, An 81.6 microW FastICA processor for epileptic seizure detection, IEEE Transactions On Biomedical Circuits And Systems 9 (1) (2015) 60–71.

[31] Varsha Harpale, Vinayak Bairagi, FPGA based architecture implementation for epileptic seizure detection using one way ANOVA and genetic algorithm, Biomedical and Pharmacology Journal 12 (3) (2019) 1543–1553.

[32] C. Chen, E. Chua, S. Tseng, C. Fu, W. Fang, Implementation of a hardware-efficient EEG processor for brain monitoring systems, in: IEEE Biomedical Conference, 2012.

[33] M. Jenihhin, M. Gorev, V. Pesonen, EEG analyzer prototype based on FPGA, in: 7th International Symposium on Image and Signal Processing and Analysis, 2010.

[34] W.M. Chen, A fully integrated 8-channel closed-loop neural prosthetic SoC for real-time epileptic seizure control, International Solid-State Circuits Conference (2013) 286–287.

[35] C.H. Yang, Y.H. Shih, H. Chiueh, An 81.6 microW FastICA processor for epileptic seizure detection, IEEE Transactions On Biomedical Circuits And Systems 9 (1) (2015) 31–47.

[36] Jui-Chieh Liao, Wai-Chi Fang, An ICA-based automatic eye blink artifact eliminator for real-time multi-channel EEG applications, in: IEEE International Conference on Consumer Electronics (ICCE), 2013, pp. 532–535.

[37] M.H. Soomro, N. Badruddin, M.Z. Yusoff, M.A. Jatoi, Automatic eye-blink artifact removal method based on EMD-CCA, in: International Conference on Complex Medical Engineering (CME), 2013, pp. 186–190.

[38] S. Jirayucharoensak, P. Israsena, S. Pan-ngum, S. Hemrungrojn, Online EEG artifact suppression for neurofeedback training systems, in: 6th International Conference (BMEiCON) on Biomedical Engineering, 2013, pp. 1–5.

[39] J-C. Liao, W-Y. Shih, K-J. Huang, W-C. Fang, An online recursive ICA based real-time multichannel EEG system on chip design with automatic eye blink artifact rejection, in: International Symposium on VLSI Design, Automation, and Test, 2013, pp. 1–4.

[40] C. Bedoya, D. Estrada, S. Trujillo, N. Trujillo, D. Pineda, J.D. Lopez, Automatic component rejection based on fuzzy clustering for noise reduction in electroencephalographic signals, in: XVIII Symposium of Image, Signal Processing, and Artificial Vision (STSIVA), 2013, pp. 1–5.

[41] A. Zachariah, Jinu Jai, G. Titus, Automatic EEG artifact removal by independent component analysis using critical EEG rhythms, in: International Conference on Control Communication and Computing (ICCC), 2013, pp. 364–367.
[42] N. Mammone, F. La Foresta, F.C. Morabito, Automatic artifact rejection from multichannel scalp EEG by wavelet ICA, IEEE Journal of Sensors (2012) 533–542.
[43] U. Heute, A.S. Guzman, Removing cleaned eye-blinking artifacts from EEG measurements, in: International Conference on Signal Processing and Integrated Networks (SPIN), 2014, pp. 576–580.
[44] R. Mahajan, B.I. Morshed, Unsupervised eye blink artifact de-noising of EEG data with modified multi-scale sample entropy, kurtosis, and wavelet-ICA, IEEE Journal of Biomedical and Health Informatics 19 (1) (2015) 158–165.
[45] W. Chen, H. Chiueh, T. Chen, C. Ho Chi Jeng, M. Ker, C. Lin, Y. Huang, C. Chou, T. Fan, M. Cheng, Y. Hsin, S. Liang, Y. Wang, F. Shaw, Y. Huang, C. Yang, C. Wu, A fully integrated 8-channel closed-loop neural-prosthetic CMOS SoC for real-time epileptic seizure control, IEEE Journal Of Solid-State Circuits 49 (1) (2014) 12–25.
[46] W.C. Fang, C.K. Chen, E. Chua, C.C. Fu, S.Y. Tseng, S. Kang, A low power biomedical signal processing system-on-chip design for portable brain-heart monitoring systems, in: Proceeding of IEEE Biomedical Conference, 2010.
[47] K. Shyu, M. Li, FPGA implementation of FastICA based on floating-point arithmetic design for real-time blind source separation, in: Proceeding of International Joint Conference on Neural Networks, Vancouver, BC, Canada, 2006.
[48] W. Huang, S. Hung, J. Chung, M. Chang, L. Van, C. Lin, FPGA implementation of 4-channel ICA for on-line EEG signal separation, in: Biomedical Circuits and Systems Conference, 2008, pp. 65–68.
[49] T-J. Chen, Herming Chiueh, Sheng-Fu Chang, Shun-Ting Jeng, Chi Hsu, Yu-Cheng Chien Tzu-Chieh, The implementation of a low-power biomedical signal processor for real-time epileptic seizure detection on absence animal models, IEEE Journal on Emerging and Selected Topics in Circuits and Systems (2011) 21–32.
[50] T.-J. Chen, S.-C. Lee, C.-H. Yang, C.-F. Chiu, H. Chiueh, A 28.6 microwatt mixed-signal processor for epileptic seizure detection, in: Proc. Symp. VLSI Circuits, Kyoto, Japan, 2013, pp. C52–C53.
[51] M. Shoaib, K.H. Lee, N.K. Jha, N. Verma, A 0.6–107 microwatt energy-scalable processor for directly analyzing compressively-sensed EEG, IEEE Transactions on Circuits and Systems I: Regular Papers 61 (4) (2014) 1105–1118.
[52] Kazutaka Jin, Long-cherished dreams for epileptologists and clinical neurophysiologists: automatic seizure detection in long-term scalp EEG, Journal of Clinical Neurophysiology (2014).
[53] Kai Fu, Jianfeng Qu, Yi Chai, Yong Dong, Classification of seizure based on the time-frequency image of EEG signals using HHT and SVM, Journal of Biomedical Signal Processing and Control 13 (2014) 15–22.
[54] Serkan Kiranyaz, Turker Ince, Morteza Zabihi, Dilek Ince, Automated patient-specific classification of long-term electroencephalography, Journal of Biomedical Informatics 49 (2014) 16–31.
[55] Mohammad Zavid Parvez, Manoranjan Paul, Epileptic seizure detection by analyzing EEG signals using different transformation techniques, Journal of Neurocomputing 145 (5) (2014) 190–200.
[56] Z. Zainuddin, L.K. Huong, O. Pauline, Reliable epileptic seizure detection using an improved wavelet neural network, Australasian Medical Journal (AMJ) 6 (5) (2013) 308–314.
[57] W.J. Freeman, Combining fMRI with EEG and MEG in order to relate patterns of brain activity to cognition, International Journal of Psychophysiology 73 (1) (2009) 43–52, https://doi.org/10.1016/j.ijpsycho.2008.12.019, Epub 2009 Feb. 20.

[58] Lan-Lan Chen, Jian Zhang, Jun-Zhong Zou, Chen-Jie Zhao, Gui-Song Wang, A framework on wavelet-based nonlinear features and extreme learning machine for epileptic seizure detection, Journal of Biomedical Signal Processing and Control 10 (2014) 1–10.
[59] Ram Bilas Pachori, Shivnarayan Patidar, Epileptic seizure classification in EEG signals using second-order difference plot of intrinsic mode functions, Journal of Computer Methods and Programs in Biomedicine 113 (2) (2014) 494–502.
[60] Sang-Hong Lee, Joon S. Lim, Jae-Kwon Kim, Junggi Yang, Youngho Lee, Classification of normal and epileptic seizure EEG signals using wavelet transform, phase-space reconstruction, and Euclidean distance, Journal of Computer Methods and Programs in Biomedicine 116 (1) (2014) 10–25.
[61] Yatindra Kumar, M.L. Dewal, R.S. Anand, Epileptic seizure detection using DWT based fuzzy approximate entropy and support vector machine, Journal of Neurocomputing 133 (10) (2014) 271–279.
[62] Varun Joshi, Ram Bilas Pachori, Antony Vijesh, Classification of ictal and seizure-free EEG signals using fractional linear prediction, Journal of Biomedical Signal Processing and Control 9 (2014) 1–5.
[63] Peng Xu, Xiuchun Xiong, Qing Xue, Peiyang Li, Rui Zhang, Zhenyu Wang, Pedro A. Valdes-Sosa, Yuping Wang, Dezhong Yao, Differentiating between psychogenic nonepileptic seizures and epilepsy based on common spatial pattern of weighted EEG, Resting Networks, IEEE Transactions On Biomedical Engineering 61 (6) (2014) 1747–1755.
[64] Information sheet, Non-epileptic Seizures, BC Epilepsy Society, Approved for circulation, 2005.
[65] S.R. Benbadis, L. Heriaud, Psychogenic (non-epileptic) seizures, a guide for families and patients, http://hsc.usf.edu/COM/epilepsy/PNESbrochure.pdf.
[66] Information for Patients, Non-Epileptic Seizures: a Short Guide for Patients and Families, Department of Neurology Royal Hallamshire Hospital.
[67] J. Kuyk, F. Leijten, H. Meinardi, Ph. Spinhoven, R. Van Dyck, The diagnosis of psychogenic non-epileptic seizures: a review, Seizure 6 (4) (1997) 243–253.
[68] E. Pippa, et al., Classification of epileptic and non-epileptic EEG events, in: Wireless Mobile Communication and Healthcare (Mobihealth), 2014 EAI 4th International Conference on, Athens, 2014, pp. 87–90.
[69] K. Turner, A. Piazzini, V. Chiesa, V. Barbieri, A. Vignoli, E. Gardella, G. Tisi, S. Scarone, M.P. Canevini, O. Gambini, Patients with epilepsy and patients with psychogenic non-epileptic seizures: video-EEG, clinical and neuropsychological evaluation, Seizure 20 (9) (2011) 706–710.
[70] J.A. Pillai, S.R. Haut, Patients with epilepsy and psychogenic non-epileptic seizures: an inpatient video-EEG monitoring study, Seizure 21 (1) (2012) 24–27.
[71] J. Durrant, H. Rickards, A.E. Cavanna, Prognosis and outcome predictors in psychogenic nonepileptic seizures, Epilepsy Research and Treatment 7 (2011), https://doi.org/10.1155/2011/274736, Epub 2011 Feb. 9.
[72] J. Gubbi, S. Kusmakar, A.S. Rao, B. Yan, T. O'Brien, M. Palaniswami, Automatic detection and classification of convulsive psychogenic nonepileptic seizures using a wearable device, IEEE Journal of Biomedical and Health Informatics 20 (4) (2016) 1061–1072.
[73] J. Bayly, J. Carino, S. Petrovski, M. Smit, D.A. Fernando, A. Vinton, B. Yan, J.R. Gubbi, M.S. Palaniswami, T.J. O'Brien, Time-frequency mapping of the rhythmic limb movements distinguishes convulsive epileptic from psychogenic nonepileptic seizures, Epilepsia 54 (2014) 1402–1408.
[74] I.Kotsopoulos, M. de Krom, F. Kessels, J. Lodder, J. Troost, M. Twellaar, T.V. Merode, A. Knottnerus, Incidence of epilepsy and predictive factors of epileptic and non-epileptic seizures, Seizure 14 (3) (2005) 175–182.

[75] A. Ponnusamy, J.L.B. Marques, M. Reuber, Comparison of heart rate variability parameters during complex partial seizures and psychogenic nonepileptic seizures, Epilepsia 53 (2012) 1314–1321.
[76] Christophe C. Jouny, Piotr J. Franaszczuk, Gregory K. Bergey, Improving early seizure detection, Epilepsy & Behavior 22 (l1) (2011) S44–S48.
[77] Sriram Ramgopal, Sigride Thome-Souza, Michele Jackson, Navah Ester Kadish, Iván Sánchez Fernández, Jacquelyn Klehm, William Bosl, Claus Reinsberger, Steven Schachter, Tobias Loddenkemper, Seizure detection, seizure prediction, and closed-loop warning systems in epilepsy, Epilepsy and Behavior 37 (2014) 291–307.
[78] Ali Yadollahpour, Mustafa Jalilifar, Seizure prediction methods: a review of the current predicting techniques, Biomedical and Pharmacology Journal 7 (1) (2014).
[79] Nilufer Ozdemir, Esen Yildirim, Patient specific seizure prediction system using Hilbert spectrum and Bayesian networks classifiers, Computational and Mathematical Methods in Medicine 2014 (2014) 572082, https://doi.org/10.1155/2014/572082.
[80] Arthur Petrosian, Danil Prokhorov, Richard Homan, Richard Dasheiff, Donald Wunsch II, Recurrent neural network based prediction of epileptic seizures in intra- and extracranial EEG, Neurocomputing 30 (2000) 201–218.
[81] B. Direito, F. Ventura, C. Teixeira, A. Dourado, Optimized feature subsets for epileptic seizure prediction studies, in: Engineering in Medicine and Biology Society, EMBC, 2011 Annual International Conference of the, IEEE, Boston, MA, 2011, pp. 1636–1639.
[82] C.Y. Chiang, N.F. Chang, T.C. Chen, H.H. Chen, L.G. Chen, Seizure prediction based on classification of EEG synchronization patterns with on-line retraining and post-processing scheme, in: Engineering in Medicine and Biology Society, EMBC, 2011 Annual International Conference of the, IEEE, Boston, MA, 2011, pp. 7564–7569.
[83] K.A. Helini Kulasuriya, M.U.S. Perera, Forecasting epileptic seizures using EEG signals, wavelet transform and artificial neural networks, in: IT in Medicine and Education (ITME), 2011 International Symposium on, Cuangzhou, 2011, pp. 557–562.
[84] James R. Williamson, Daniel W. Bliss, David W. Browne, Jaishree T. Narayanan, Seizure prediction using EEG spatiotemporal correlation structure, Epilepsy and Behavior 25 (2) (2012) 230–238.
[85] C. Teixeira Bruno Direito, Mojtaba Bandarabadi, Hans Peter Grebe, Francisca S. Francisco Sales, Antonio Dourado, Real-time epileptic seizure prediction at Centro Hospitalar e Universitário de Coimbra, in: Experiment International Conference (exp.at'13), Coimbra, 2013, pp. 96–198.
[86] Sergi Consul, Bashir I. Morshed, Robert Kozma, Hardware efficient seizure prediction algorithm, in: Proc. SPIE 8691, Nanosensors, Biosensors, and Info-Tech Sensors and Systems 2013, 2013, p. 86911J.
[87] A. Sharma, J.K. Rai, R.P. Tewari, Epileptic seizure prediction and identification of epileptogenic region using EEG signal, in: Green Computing and Internet of Things (ICGCIoT), International Conference on, Noida, 2015, pp. 1194–1198.
[88] M.Z. Parvez, M. Paul, Epileptic seizure prediction by exploiting spatiotemporal relationship of EEG signals using phase correlation, IEEE Transactions on Neural Systems and Rehabilitation Engineering 24 (1) (2016) 158–168.
[89] L.J. Bonnett, C. Tudur-Smith, P.R. Williamson, A.G. Marson, Risk of recurrence after a first seizure and implications for driving: further analysis of the multicentre study of early epilepsy and single seizures, BMJ. British Medical Journal (2010) 341–362.
[90] R. Sukanesh, R. Harikumar, Minimum relative entropy (MRE) method for fuzzy based classification of epilepsy risk levels from EEG signals, in: International Conference on Biomedical and Pharmaceutical Engineering, Singapore, 2006, pp. 93–98.
[91] R. Harikumar, M. Balasubramani, T. Vijayakumar, Performance analysis of patient specific epilepsy risk level classifications from EEG signals using two tier hybrid (fuzzy,

soft decision trees models and MLP neural networks) classifiers, International Journal of Soft Computing and Engineering (IJSCE) 2 (2) (2012) 541–549.
[92] J.A. Cramer, J. French, Quantitative assessment of seizure severity for clinical trials: a review of approaches to seizure components, Epilepsia 42 (1) (2001) 119–129.
[93] F. Yaghouby, B. Madahian, H. Mirinejad, S. Sunderam, Feasibility of seizure risk prediction using intracranial EEG measurements in dogs, in: 2015 41st Annual Northeast Biomedical Engineering Conference (NEBEC), 2015, pp. 1–2.
[94] B.K. Tripathy, Sasikumar Gurumurthy, Classification and analysis of EEG brain signals for finding epilepsy risk levels using SVM, World Applied Sciences Journal 33 (2015) 631–639.
[95] V.K. Harpale, V.K. Bairagi, Significance of independent component analysis (ICA) for epileptic seizure detection using EEG signals, in: Proceedings of the International Conference on Data Engineering and Communication Technology (ICCUBEA2017), Pune, in Advances in Intelligent Systems and Computing, vol. 2, 2017, pp. 829–838.
[96] V.K. Harpale, V.K. Bairagi, An adaptive method for feature selection and extraction for classification of epileptic EEG signal in significant states, Journal of King Saud University – Computer and Information Sciences (ISSN 1319-1578) (2018), https://doi.org/10.1016/j.jksuci.2018.04.014.

CHAPTER 2

Concepts of EEG analysis and seizures

2.1 EEG analysis and seizure detection methodology

This book focused on the early prediction of seizures, seizure detection, and classification of seizures as epileptic and nonepileptic with improved classification accuracy. The scheme analyzes electroencephalography (EEG) signals for seizure detection and prediction. The research methodology is divided into signal processing, feature extraction, detection of seizures, and classification in epileptic and nonepileptic seizures. It can be implemented, as shown in Fig. 2.1. This methodology also can be upgraded to a wireless seizure monitoring system which will help doctors to provide a solution to epileptic patients in rural areas. The EEG signal is recorded with multiple electrodes referred to as channels. If some of the channels have variation during the seizure, this is called a partial seizure, and if all channels are affected by the seizure state, this is called a generalized seizure.

Fig. 2.2 shows EEG signals recorded by 23 channels, of which channels 1–5, 8–12, and 20–23 show more variations in amplitude and frequency than the remaining channels. Thus the EEG signal recorded has

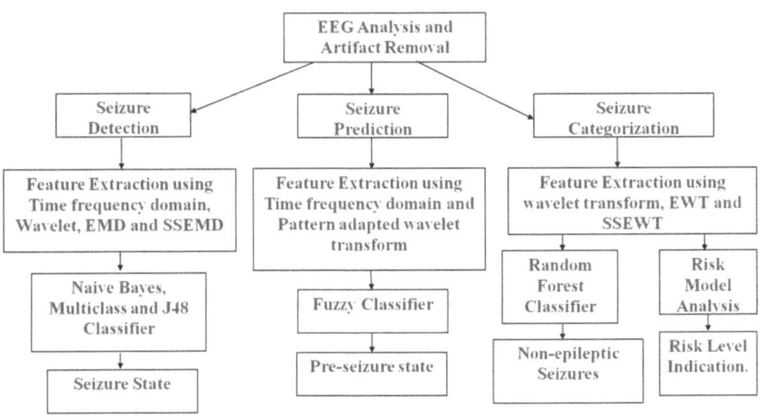

Figure 2.1 Flow diagram of research methodology.

Brain Seizure Detection and Classification Using EEG Signals
https://doi.org/10.1016/B978-0-32-391120-7.00007-4

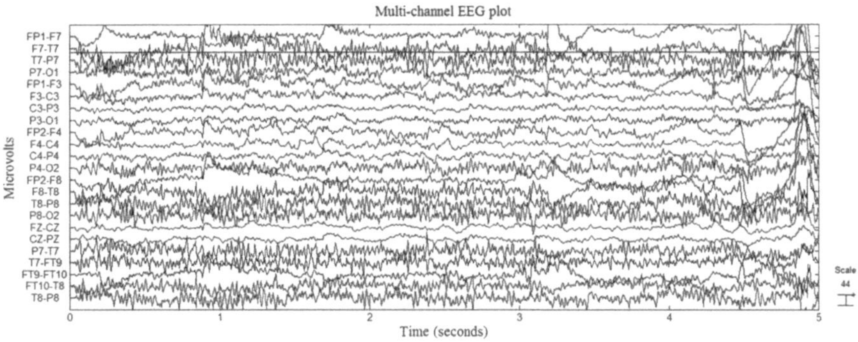

Figure 2.2 EEG signal representation of 23 channels.

partial seizure characteristics. The average potential observed with EEG is 0 to 200 μV, with a frequency range from 0.4 to 104 Hz. The rhythms characterized by EEG frequency are delta activity (0.4 to 4 Hz), theta activity (4 to 8 Hz), alpha activity (8 to 13 Hz), beta activity (13 to 30 Hz), and gamma activity (30 to 60 Hz), as shown in Fig. 2.3. It shows wavelet decomposition of a single channel with separated frequency bands and its energy distribution.

a. Database

Initially, the Andrzejak (2012) [1] dataset was used for experimentation. It contains tonic-clonic seizure files of two subjects recorded with a scalp right central (C4) electrode with linked earlobe reference. It has a recording of 3 min with about 1 min preseizure, the seizure, and some postseizure activity. Its sampling rate is 102.4 samples/seconds; thus, the total number of samples in the file is 18,432, used with a 2Fs window of 204 samples for experimentation.

The standard EEG database described in Andrzejak (2001) [2] is a large dataset used for experimentation and evaluating proposed methodologies related to seizure detection. The dataset is comprised of five sets (A, B, C, D, and E) recorded using 128 channels with average reference and 12-bit ADC resolution. Each of the set contains 100 EEG segments with a sampling frequency of 173.60 Hz and a recording duration of 23.6 seconds. Thus, each dataset represents 100 files × 4097 samples data points for feature extractions. Normally a dataset has physically recorded signals with moving artifacts, but this database is artifact-free, thus avoiding the need for preprocessing. Set A (recorded with eyes open) and set B (eyes closed) are scalp EEG-based datasets

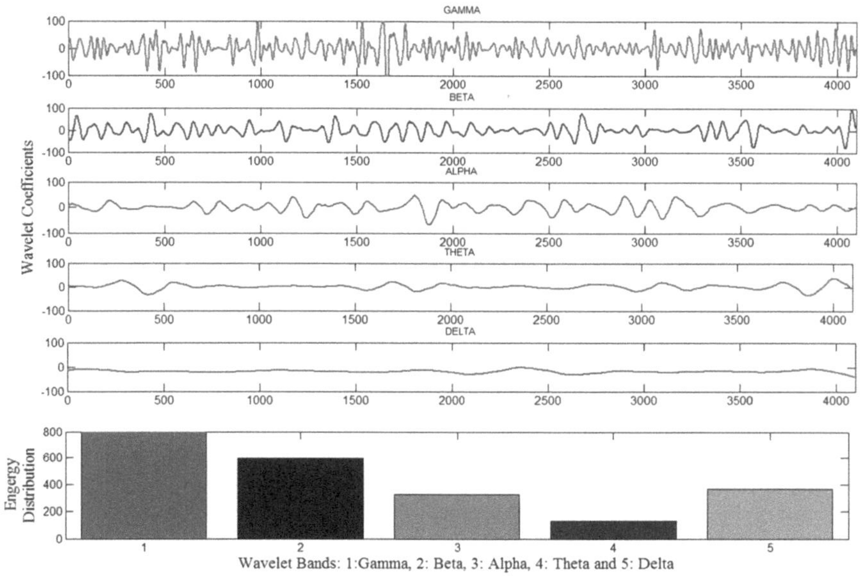

Figure 2.3 EEG signal decomposition and energy distribution in gamma, beta, alpha, theta, and delta.

of five healthy subjects, whereas sets C, D, and E are intracranially recorded signals from five epileptic patients. While sets D and C contain the EEG signals recorded during seizure-free intervals from the affected epileptic hemisphere and the nonaffected hemisphere of the brain, respectively, set E contains the onset seizure signal. Experimentation sets A (eyes open), B (eyes closed), and E are used for classifying EEG signals in three classes, Normal_o, Normal_c, and Seizure, respectively.

The CHB-MIT EEG scalp dataset from the Children's Hospital at Boston [3] is used for this experimentation. It is an epileptic seizure-based database with scalp EEG recordings from 22 patients. It is comprised of 5 males and 17 females with ages ranging from 1.5 to 22 years. The recorded EEG data have a sampling rate of 256 samples per second with 16-bit resolution. The international 10-20 system bipolar was used for location of sensors with 23 channels to record the data. The files are available in chb01 to chb22 sets. Each set contains more than 20 EDF files with 1–4-hour recordings, thus the maximum number of samples is 4 × 60 × 256 (sampling rate), i.e., one file with maximum 8,064,000 samples and minimum 2,016,000 samples with 23 channels.

Table 2.1 List of seizure files in the TUH database.

Seizure type	Training event files	Testing event files
FNSZ	663	284
GNSZ	129	187
SPSZ	48	6
CPSZ	320	92
ABSZ	52	42
TNSZ	17	50
TCSZ	25	17
MYSZ	0	1
Nonepileptic	145	89
Normal	590	247
Total files	1989	1015

For the classification method the TUH EEG Seizure Corpus (TUSZ) [4] (American Clinical Neurophysiology Society Phoenix, Arizona, USA) dataset is used. This dataset is significantly larger than previous datasets. The dataset is labeled dataset and was recorded using various techniques, locations, and sampling rates (mostly 250 and 400 Hz). It is comprised of two parts, training and testing, with 264 subjects and 50 subjects, respectively. The files handled for experimentation are 1254 files of training and 273 records from the testing category in European Data Format (EDF format). The TUSC database contains seizure files of various seizure types such as tonic-clonic seizures, myoclonic seizures, and partial seizures, along with nonepileptic EDF files, as described in Table 2.1.

The EEG signals were recorded using the epilepsy monitoring unit with 32 channels. This dataset contains some data recorded with the scalp EEG recording method and other data recorded with the intracranial method. The number of files used for experimentation areas are listed in Table 2.1. Thus, a standard labeled database was used for experimentation, and results were verified and validated by a neurologist.

b. EEG sensors

EEGs are recorded with scalp EEG using electrodes such as Agcl scalp reusable electrodes, EEG cap, and adhesive gel electrodes. For intracranial EEG recordings, subdermal needles are inserted in the first layer of the brain. Intracranial recordings are typically used in complex surgeries. As proposed research work is patient-specific, a previously recorded database by various neurologists is used to prove the concept. An available small dataset and a large dataset of epilepsy seizures are used for initial experimentation. The design of a real-time brain signal monitoring system in this work will facilitate the development of portable or wireless seizure detection devices.

c. Preprocessing

EEG/MEG recorded brain signals produced by sensors need to be processed to eliminate the influence of disturbance observed because of body interference and environmental conditions. Filtering methods such as notch filters and adaptive filters are used for removing these interferences. Also, multichannel inputs need to be separated, for which various techniques can be used, such as principal component analysis (PCA) and independent component analysis.

d. Feature extraction

For classification and identification of seizures, features need to be extracted from nonlinear EEG signals. Empirical mode decomposition, singular value decomposition (SVD), and discrete wavelet transformation are the most commonly used algorithms for the feature extraction of EEG signals. The fundamental part of the Hilbert–Huang transformation [5] is the EMD method, in which any linear/nonlinear dataset is represented as a finite and small number of components called intrinsic mode functions [6]. SVD is a very well-known method in linear algebra and has been extensively applied to signal processing, statistical analysis, and mathematical modeling.

e. Classification of seizures

The objective of the work is the detection and prediction of seizures. The first step is the classification of EEG signals into a seizure (ictal state) and nonseizure (interictal) signals. In recent research work, various classifiers were proposed, such as particle swarm optimization (PSO), conditional mutual information maximization (CMIM) [15], genetic algorithm with SVM classifier, and many more specified in the literature. The primary objective of classification is to compare all these methods based on performance measurement and propose an advanced algorithm to enhance the performance. The classified seizures are also processed to classify them into epileptic and nonepileptic (PNES) seizures, and then risk analysis will be applied to diagnose PNES accurately.

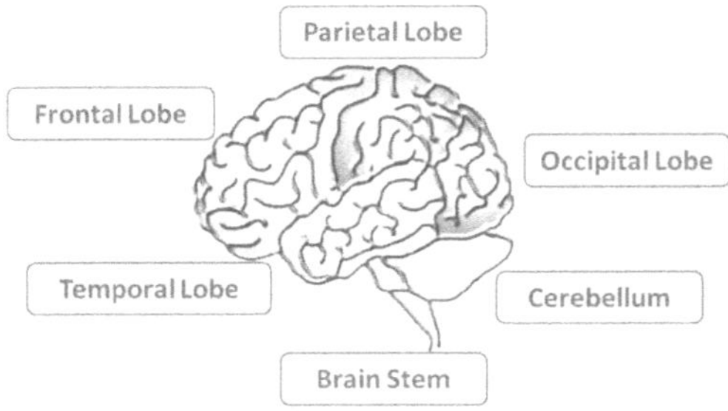

Figure 2.4 Functional parts of the cerebral cortex [7].

f. **Prediction of seizures**
Prediction of seizures is urgently needed as it is more advantageous to predict than to detect seizures. It will avoid many accidents and aid clinicians to provide preventive care for the patient. Various methods have been developed to design a prediction model, but challenges persist due to high false positive rates.

g. **Risk analysis**
In the field of risk analysis of seizures, various parameters are considered for risk analysis. Few parameters are quantitative, and the remaining are technical parameters. Risk assessment is performed with the quality value, which is observed to be as low as 22.23%. A risk model based on technical characteristics of detected seizures was proposed and demonstrated either by a real-time system or software such as MATLAB®.

2.2 Brain signal behavior, measurement, and recording

The brain is a system composed of billions of interconnected neurons, and it is an essential part of the central nervous system. The brain is functionally divided into the cerebral cortex and thalamus, among others, where the cerebral cortex is a thin layer of 2–3 mm that covers the entire brain and the thalamus is a medium of communication between the subcortical region of the brain and the cortex for further activity control. The cerebral cortex controls various cognitive and motor actions of the human body. Fig. 2.4 shows the functional parts of the brain, namely, the temporal lobe, the parietal lobe, the occipital lobe, and the frontal lobe. The behavior of the brain is classified as microscopic, mesoscopic, and macroscopic based on its spatial scale. The microscopic scale describe small-scale spatial behavior; it includes a single or a few brain cells. Macroscopic behavior is observed at large spatial scales (cm), covering whole regions of the brain. Mesoscopic behavior is a response of networks of neurons with a spatial scale of millimeters. The temporal scales of the brain or behavior are expressed based on the activity frequency of the neurons. Because of fast mechanisms in single neurons, the microscopic scale is typically defined with frequencies of 1000 Hz and above. The mesoscopic scale is defined as frequencies between 10 and 1000 Hz. The macroscopic level is defined with frequencies in the range of 1 to 100 Hz [7]. Mathematical brain models include parameters such as

$$f(brain) = (input,\ internal\ signals,\ delays,\ and\ outputs). \tag{2.1}$$

Let the model of the microscopic model for a single neuron be given as

f_{micro} (brain) = activity of synapses, integrations of postsynaptic potential (PSP) at soma, the propagation time of information from soma to synapses, action potential.

This mathematical model supports the generation of event-related potentials (ERPs). ERPs are a component of the EEG signal. Analysis of the normal and abnormal behavior of the brain is possible by evaluating this model for healthy and affected brains. The EEG signal is an average rate of change of ERPs per seconds observed due to excitation of neurons covered under a single electrode.

2.3 EEG processing and mathematical model

EEG is based on electrical variation produced by the firing of neurons within the brain along the scalp. For the recording of EEGs, electrodes are placed at some key points on the patient's head. The 10-20 location standard is the internationally recommended method for placement of electrodes for EEG recording. The 10-20 indicates that the distance between succeeding electrodes is either 10% or 20%. Fig. 2.5 shows a location map for 23 EEG channels plotted with the EEGLAB tool. It shows the contribution of 10 and 20 in the distribution of electrodes on the scalp as described by Trans Cranial Technologies [8]. The electrodes are named by lobes of the brain; for example, "F" stands for frontal lobe and "T" for temporal lobe, and similarly for the occipital and parietal lobes. The number shows the electrode location on the hemisphere. Thus, even numbers (2, 4, 6, 8) indicate the right hemisphere and odd number the left hemisphere. The brain signal recording system can be modeled, as shown in Fig. 2.6. It is

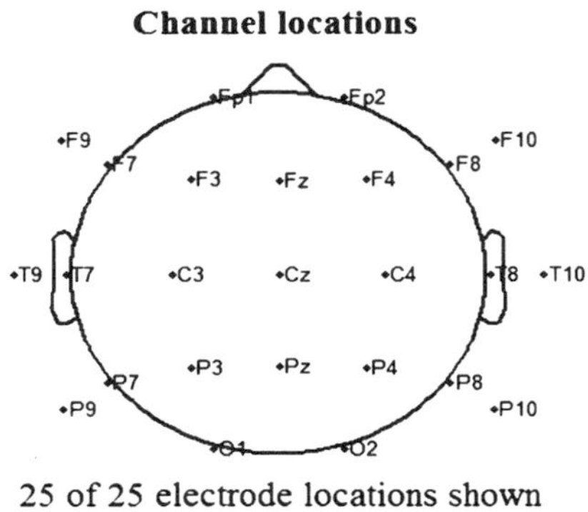

Figure 2.5 Location map with 10/20 electrode positioning system by the EEGLAB tool.

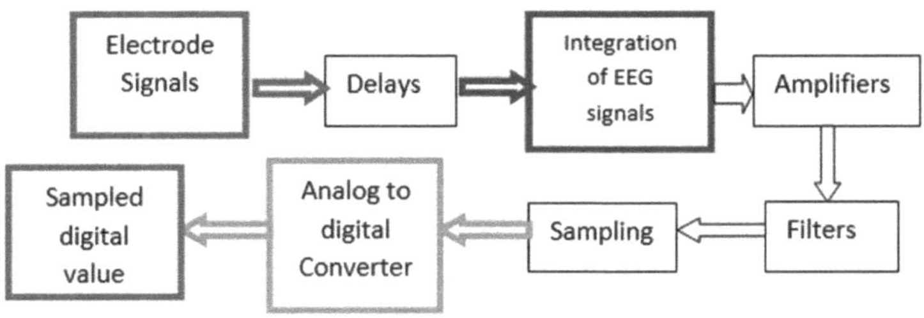

Figure 2.6 Measurement and recording of brain signals.

comprised of electrode signals as an input to the system. These input signals are integrated after some delay and recorded by the existing EEG recording system. The preprocessing EEG signals include amplification, filtration, and digital conversion to produce a sampled output with a specific resolution.

EEG is based on electric variation observed in the brain by chemical imbalance resulting in a change in voltage. It has a sufficient temporal resolution to record the fast changes in electrical activity, but its spatial resolution is poor because of limited electrodes, their localization, and properties of the head.

In existing EEG recording systems, channel synchronization and artifacts are major limitations. Mahdi Jalili et al. [9] review the significance of multivariate synchronization analysis as compared to other synchronization methods used for EEG analysis and channel synchronization.

2.3.1 Types of EEG based on placement of electrodes

1. **Scalp EEG**

In the scalp EEG method, the electrodes are placed evenly on the scalp, which helps to record brain activity. Maximally 256 electrodes can be uniformly localized, synchronized, and placed on the scalp. The brain signals are affected by various layers of cerebrospinal fluid, skull, and scalp. A normal adult scalp EEG signal is about 10 to 100 μV and is about 10–20 mV if recorded from subdural electrodes. Scalp EEG is a simple and cheap information-rich tool used for patient monitoring. Some of the characteristics of these signals are:

- The signals are highly attenuated.
- The 10-20 placement system is used to place electrodes.
- Easy and cheap, with a small setup compared to intracranial EEG.
- It is typically used for monitoring of patients.

2. **Intracranial EEG**

Intracranial EEGs are recorded by inserting electrodes on the cortex called cortical or depth electrodes, which usually penetrate the subcortical systems. The macroscopic or mesoscopic model is used to define the spatial characteristics of intracranial EEG. Intracranial records are useful for presurgical analysis. The characteristics of intracranial EEG signal are:

- No fixed placement standard is used for electrode location, as it depends on the patient's disability or brain injury.
- It is complex and expensive, with a big recording setup compared to scalp EEG.
- It is used for presurgical analysis and during surgery.

The method explained uses scalp EEGs from the CHB-MIT scalp database, and intracranial EEGs are also used from the Andrzejak (2001) database.

2.3.2 Mathematical model of EEG signals

The mathematical model describes the static and dynamic behavior of the EEG signal in terms of various parameters. The EEG signal is a continuous-time signal. Temporal characteristics of EEG signals at different locations are identified by various channels for continuous time "t." If sampled with a sampling time T, EEG signal over the continuous time t is modeled as

$$X[m] = X(mT);\ m = 1;\ 2;\ 3.$$

Here X[m] is the discrete time signal sampled at time intervals t = T;2T; ....; mT. The behavior of the signal over the total time is called the dynamics of signal z[m]. This state signifies $z[n] \in \Re^N$, where $\Re^N$ is the phase space of the system. X[m] is a discrete time measured from the single

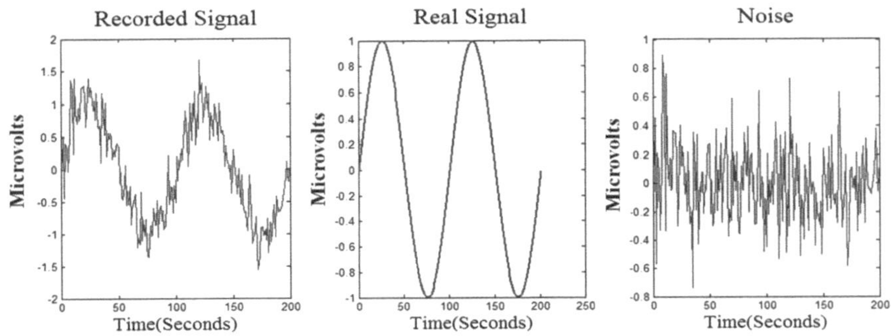

Figure 2.7 Recorded signal decomposed in real signal and noise.

channel behavior of the brain. Mathematically the resultant behavior is a summation of the true signal and added noise to it: $X[m] = X(mT) + \eta(mT)$; $m = 1; 2; 3......$
Here X (mT) is the value of the true signal at time $t = mT$ (where T is the sampling interval) and η (mT) is the error of this measurement at this time. If η (mT) is small, then $X(mT) \approx X[m]$.

Real signals can be separated from the measured EEG signal with suitable filtering, as shown in Fig. 2.7. Jun Lu et al. [10] proposed an adaptive spatiotemporal filtering method to avoid these noisy characteristics of the EEG signal. It also optimizes temporal information more accurately in lower-dimensional spaces.

Thus the quantitative analysis of EEG signals includes measurement of the mean spectral magnitude or power for multiple frequency bands. Quantitative analysis is the process of measuring specific properties of raw EEG signals and transforming these into digitally recorded EEG signals. Numerical parameters are analyzed for various applications such as brain disorder detection and emotion analysis.

The MATLAB toolbox EEGLAB is used for artifact and noise removal. Harpale et al. [11] demonstrated the use of the EEGLAB toolbox and the significance of ICA in removing artifacts. Each EEG signal is processed with the EEGLAB toolbox to remove this powerline noise, as shown in Fig. 2.8. An FIR band stop filter is designed and applied to each of the channels by the tool. It shows a suitable filter response for removal of 60-Hz powerline noise. The experimentation is performed for fl = 45 Hz and fh = 65 Hz and a band stop filter of the order 10, 15, and 18. It is observed that filter order 18 gives better noise rejection. Normally all the artifacts are in the range of 45 to 60 Hz.

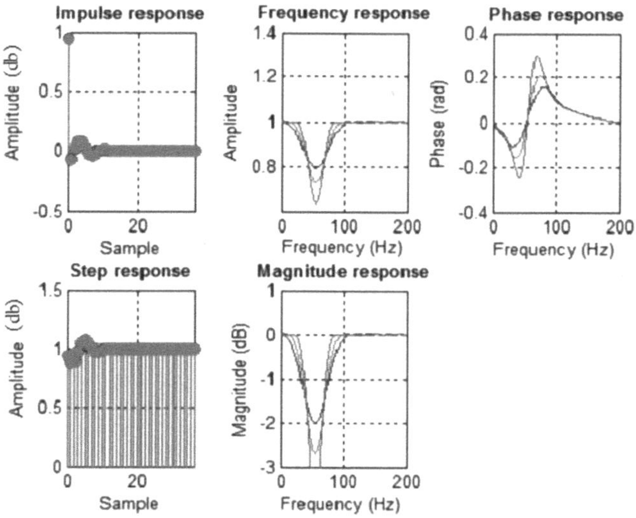

Figure 2.8 FIR filter response for order of 10, 15, and 18.

2.4 Artifacts observed in EEG signals

Due to the very low amplitude of EEG signals, they are easily contaminated by external sources. These artifacts, inherent to scalp EEG recordings, are produced by head movements, electrocardiography measurements, and muscle activity. Artifacts are unavoidable electrical signals observed during scalp recording and originate from noncerebral areas. They are caused by some external source such as electromagnetic interference or improper electrode–scalp junctions or some biological sources such as eye blinks, eye movements, extraocular muscle activity, chewing, muscle movement of the scalp, and sometimes heartbeats. Various strategies exist which deal with artifacts, such as:

- Ignoring: Some feature extraction methods consider artifact effects negligible.
- Rejecting: EEG channel information having the same significance on measurements eventhough artifacts are excluded from further analysis.
- Removing: Artifacts are filtered or separated from the actual EEG signal, and only valid signals are used for analysis. Loss of data due to rejection is very little.
- Training: A training-based mechanism is developed to identify some common artifacts and remove them.

One crucial point is that artifacts usually limit the length of EEG recordings that can be considered as stationary. Hartmann et al. [12] proposed an artifact removal algorithm to effectively remove artifacts from EEGs.

EEG signals are recorded with the multichannel acquisition module. It is very data-intense and time- and resource-consuming because it should handle a heavy workload of computations. As EEG signals are composed of various random signals, independent component analysis (ICA) is considered to be a significant method. This work specifically studies the significance of ICA in epileptic seizure detection using EEG signals. ICA is mostly used for EEG artifact removal from raw EEG signals. Thus ICA is useful in removing artifacts and improving epileptic seizure detection accuracy [13].

EEGLAB is a MATLAB toolbox made available by MathWorks for processing brain signal data processed by EEG, magnetoencephalography, and any other electrophysiological method. Along with the primary processing, EEGLAB implements various analysis algorithms, artifact rejection approaches, and data visualization methods. EEGLAB supports multiple file formats and allows users to group data from several subjects and to cluster their independent components. EEG-related time-frequency features such as event-related shifts in the power spectrum of event-related spectral perturbation (ERSP) and intertrial coherence (ITC) (event-related phase-locking) can be used, as shown in Fig. 2.9. It shows an energy distribution from 5 to 25 Hz of frequency bands for the Fp_1 channel along with its amplitude variation at specific times. The ERP observed at Fp_1 describes the frequency and phase relation also.

2.4.1 Independent component analysis

ICA is a statistical tool that is generally used for blind source separation [14]. ICA also has significant application in EEG artifact removal and feature extraction [10,11].

a. **Kurtosis for non-Gaussianity**

   The arbitrary distribution of integrated independent signals tends to a Gaussian distribution. If independent signals are mixed, then they can be separated by linearly transforming signals as non-Gaussian. Thus non-Gaussianity is very important in the calculation of ICA components. Normalized signals are thus tested for non-Gaussianity using kurtosis. If signal "x" has unit variance, its kurtosis is the fourth mo-

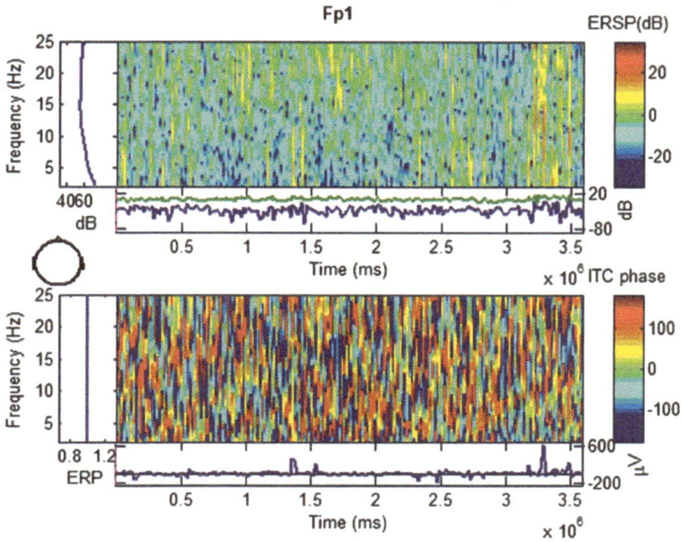

Figure 2.9 Time-frequency analysis of ERSP and ITC.

ment of that signal as defined by

$$kurt(x) = E^4 - 3(E^4)^2. \tag{2.2}$$

The independent components of one normal patient recorded with 23 channels are thus measure of non-Gaussianity of individual signals. The statistics are calculated by the EEGLAB toolbox, which shows kurtosis is nonzero and thus non-Gaussian behavior of signals, as shown in Fig. 2.10. The ordered distributions calculated from Gaussian and trimmed Gaussian distributions are shown in Fig. 2.10(a). The standard deviation observed is distributed in the first and third quadrants of the quarter quartile plot shown in Fig. 2.10(b). Thus all the statistics are individually calculated for all 23 channels. The statistics for channel FP1 show kurtosis is nonzero, thus exhibiting a non-Gaussian property.

b. Independent components

ICA is a signal processing tool that statistically separates N independent inputs which have been mixed linearly in N output channels, without knowing their dynamics. The EEG signal frequency range is the mixing of brain fields at the scalp electrodes, and their distribution is linear. ICA gives the scalp topographies of the independent components, thus supporting source localization.

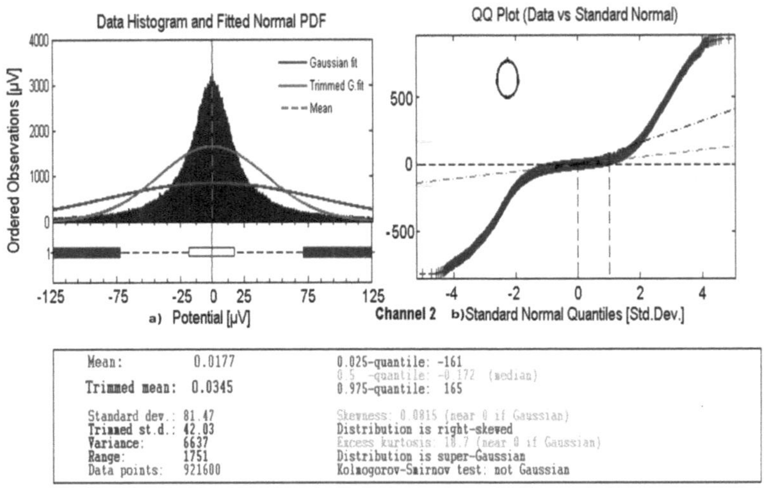

Figure 2.10 Parameters for testing non-Gaussian distributions.

2.4.2 Concept of ICA

The ICA model assumes that source signals are statistically independent and have a non-Gaussian distribution. It is composed of independent source signals *si*, an unknown mixing matrix *Ai*, and a mixed signal *x*, where $i = 1, 2, 3, \ldots, n$. This mixed signal component can be represented as

$$x = Sum(Aisi) \text{ for } i = 1 \text{ to } n. \tag{2.3}$$

Here A is assumed to be a square matrix and x and s are considered to be vectors. While obtaining independent signals, the inverse of the unknown matrix W needs to be calculated and used to find the value of s as follows:

$$s = Ai - 1x \text{ or } s = Wx. \tag{2.4}$$

This ICA has some limitations; for example, determining variance and the order of independent components is difficult. Nevertheless, it remains the preferable method for signal separation. Thus the basic objective of the ICA model is to estimate the unknown mixing matrix. Source independence can be verified with a probability density function (PDF). If the joint probability of sources $s1$ and $s2$ is $P(s1, s2)$ and the individual probability functions for $s1$ and $s2$ are $P(s1)$ and $P(s2)$, then two sources are considered to be

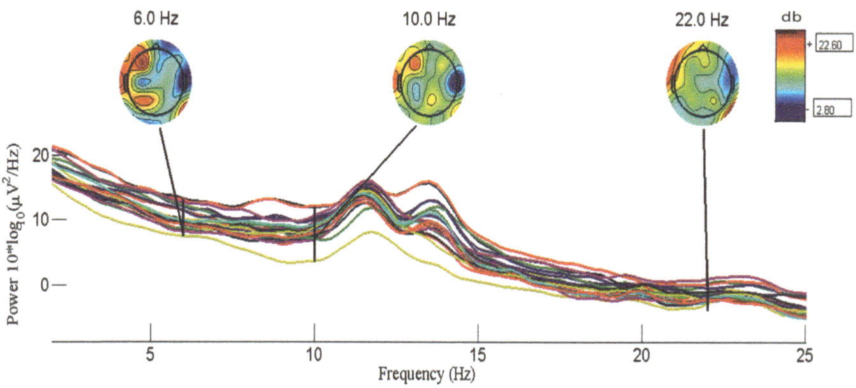

Figure 2.11 Spectro map of EEG signals.

independent only when they satisfy

$$P(s1, s2) = P(s1)P(s3). \tag{2.5}$$

Independence also can be checked with covariance of $s1$ and $s2$. Signals are said to be independent if their covariance is zero. The non-Gaussian property of signals can be verified by the kurtosis function.

2.4.3 Flow of EEGLAB tool

The work flow for removal of artifacts using ICA and EEGLAB is given as follows:

1. Steps for manual rejection of artifacts
   a. Import EDF file
   b. Import channels location file
   c. Select tools – reject continuous data
   d. Select tools – reject data epochs
2. Steps involved in automated artifact removal
   a. Import EDF file and location file
   b. Run ICA
   c. Select tool reject data using ICA
   d. Select reject by component map
   e. Select remove component
   f. Export EDF file
   g. Plot component map and properties and activation scroll.

Fig. 2.11 shows a spectro plot of all 23 channels describing frequency and power distribution relations. It shows a 2D channel map explain-

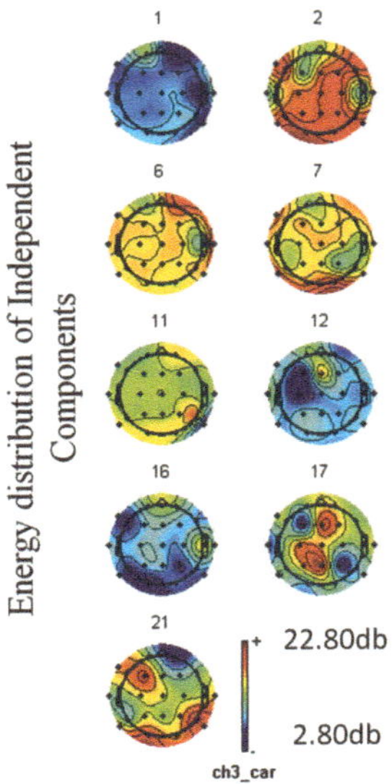

Figure 2.12 Independent components and their spatial mapping.

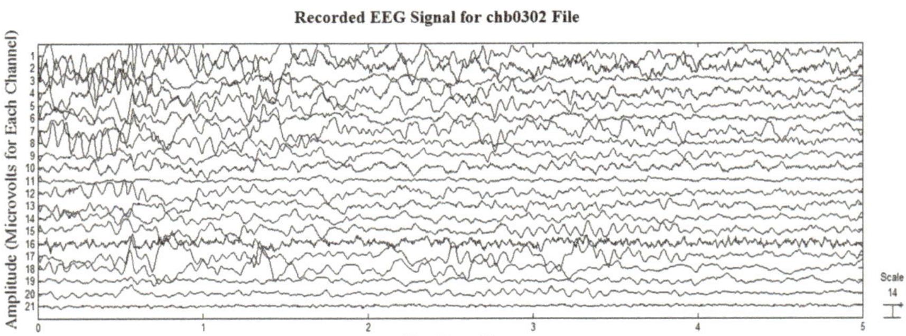

Figure 2.13 Recorded EEG signal with 23 channels.

ing channel location and the affected area near the electrode. The power distribution near the channels describes the variation in amplitude. High amplitudes show seizures that occurred near the channel.

Fig. 2.12 shows power distributions for all independent components. Channels having a high power distribution are channels by the occurrence of a seizure. This gives a clear idea about the specificity of EEG channels.

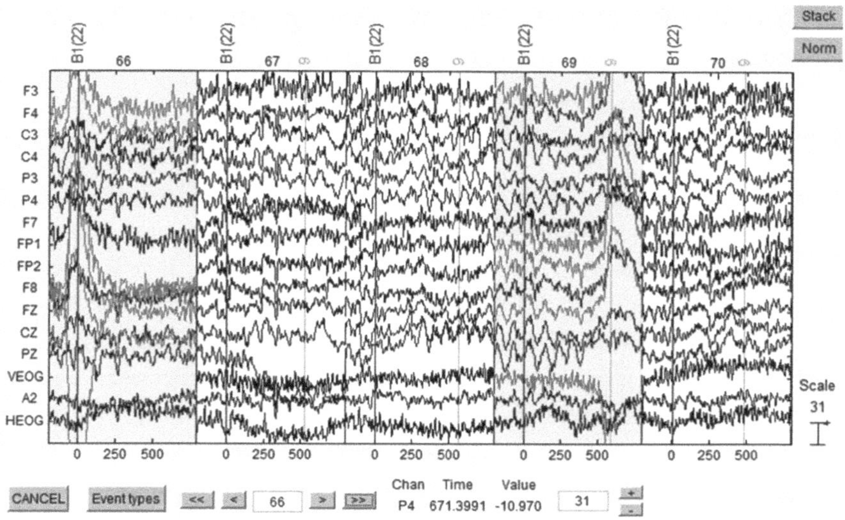

Figure 2.14 EDF file observed for manual rejection.

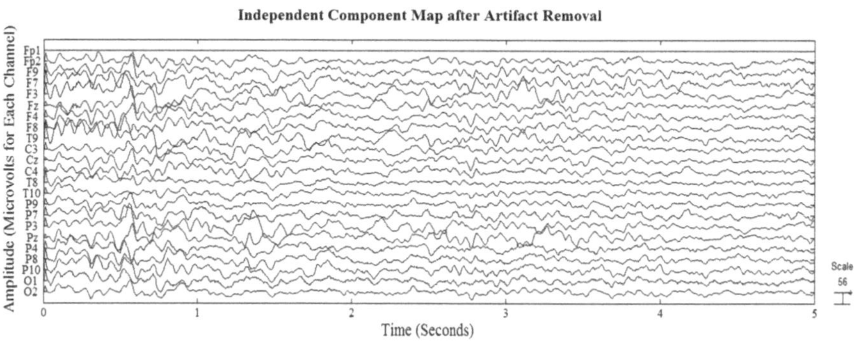

Figure 2.15 Independent components derived from EEG channels.

Figs. 2.13 to 2.15 show changes in EEG signals from real values to filtered values and finally to independent components. These signals are saved in an EDF file after removal of powerline noise and artifacts.

The input file is preprocessed and restored as the EEG signal. The artifacts rejected by this method are eye movement, eye blinking, muscle movements, and line noise. Thus each of the EDF files read in the EEGLAB tool and its independent components are extracted and artifacts are rejected manually or automatically. The preprocessed files after removal of artifacts are saved as EDF files and further used for feature extraction for abnormality detection.

2.5 Epilepsy

Epilepsy is a condition resulting from physiological abnormalities. Seizures affect patients to varying degrees and are infrequent phenomena. In seizures, neurons become hyperactive and synchronous, that is, they fire action potentials at a much higher rate than normal and at the same time, with all nearby neurons firing. Seizures are the result of an abnormality in this network that allows neural activity to become unchecked. This hyperactivity and synchronicity spreads over the brain, causing total loss of control and brain oscillations. The seizure ends because oxygen supplies to the neurons get depleted or the chemical imbalance is restored. Thus the postseizure period is also very important.

2.5.1 Basics of seizures

Seizures (i.e., ictal) are characterized by large fluctuations of the firing rate, which is caused by dysfunctional regulation of neuronal activity. Typically, temporary loss of control, abnormal movements, and unconsciousness are due to the abnormal behavior called "seizure." Nonepileptic seizures are mostly stress-related or emotional stress-induced and result from traumatic psychological experiences, sometimes from the forgotten past. They are sometimes called pseudoseizures, or "psychogenic nonepileptic seizures" (PNES). PNES or NES are normally misdiagnosed as epilepsy. With early diagnosis, seizures eventually disappear in 60–70% of adults; the percentages are even higher for children.

2.5.2 Types of seizures

According to the origin of seizures, they are described as either partial (focal) or generalized [15]. Partial seizures only involve a localized part of the brain, whereas generalized seizures involve the whole of both hemispheres. Partial seizures are by "secondary generalization," and later spread over the whole cortex and become generalized.

Seizures are broadly classified as:

1. partial or focal seizures or
2. generalized or nonfocal seizures.

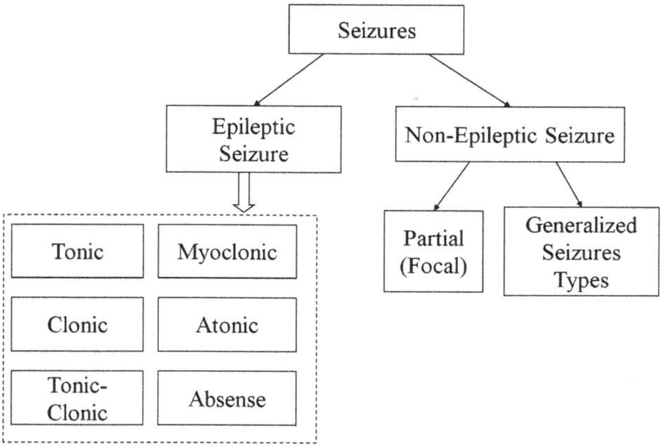

Figure 2.16 Types of seizures.

The detailed types of seizures are shown in Fig. 2.16. The seizure classes used for experimentation from the TUH database are described in Table 2.2. Seizure classification is based on these classes and nonepileptic seizure information from this database.

Focal epilepsy is a functional failure of the mesoscopic network and caused by an abnormality in a specific part of the brain in the form of damaged or abnormal neurons. When focal seizures spread they are called secondarily generalized and observed due to damaged or abnormal neurons.

Focal epilepsy can be defined as:

1. Initiation: A measure of chemical imbalance caused by increased excitability and failure of the inhibitory mechanisms, causing synchronization in abnormal neurons, which are responsible for the seizure. Understanding the initiation of a seizure may lead to the ability to predict its onset.
2. Generalization: Abnormality spreading over the brain, possibly because the thalamus becomes passive or overactive.

Focal epileptic seizures are further divided as simple and complex. A simple seizure causes mild cognitive, psychic, sensory, or autonomic symptoms and observed consciousness. Complex seizures cause convulsive behavior and unconsciousness. Nonfocal or generalized seizures are due to failure of the macroscopic neural network. They spread across the brain immediately, and the full brain is affected, which may cause unconsciousness. Some of the symptoms and causes are given below.

Table 2.2 Seizure classes from the TUH database.

Seizure code	Seizure class	Description
FNSZ	Focal nonepileptic seizure	Focal seizures which cannot be specified by its type
GNSZ	Generalized nonepileptic seizure	Generalized seizures which cannot be further classified into one of the groups below
SPSZ	Simple partial seizure	Partial seizures during consciousness
CPSZ	Complex partial seizure	Partial seizures during unconsciousness
ABSZ	Absence seizure	Absence of discharges observed on EEG; patient loses consciousness for a few seconds (petit mal)
TNSZ	Tonic seizure	Stiffening of the body during seizure (EEG effects disappear)
CNSZ	Clonic seizure	Jerking/shivering of body during seizure
TCSZ	Tonic clonic seizure	At first stiffening and then jerking of body
ATSZ	Atonic seizure	Sudden loss of muscle tone
MYSZ	Myoclonic seizure	Myoclonic jerks of limbs
NE	Nonepileptic	Nonepileptic patients without seizures
Norm	Normal	Patients not affected by neurological disorders

1. Symptoms for occurrence of a seizure disorder
   - periods of blackout or confused memory
   - occasional "fainting spells"
   - episodes of blank staring in children
   - sudden falls for no apparent reason
   - episodes of blinking or chewing at inappropriate times
2. Seizure triggers
   - Missed medication
   - Stress/anxiety
   - Dehydration
   - Lack of sleep/extreme fatigue

2.6 Summary

EEG is a noninvasive method to measure brain activity; it is simple and inexpensive and has a good temporal resolution. The disadvantages of EEG are that the signal-to-noise ratio is poor, thus extensive data analysis is required, the spatial resolution is poor, and large subject-specific databases are required. Frequency analysis of EEG signals extracts quantitative parameters and performs comparisons between the power at different frequency bands and their topological distribution. Time-frequency analysis is used in various applications such as automatic seizure detection. EEG is the preferable diagnosis tool for analyzing and detecting various brain disorders. Preprocessing of EEG signals from specified databases will be covered in the further part of the book.

References

[1] R.G. Andrzejak, K. Schindler, C. Rummel, Nonrandomness, nonlinear dependence, and nonstationarity of electroencephalographic recordings from epilepsy patients, Physical Review E 86 (2012) 046206.

[2] R.G. Andrzejak, K. Lehnertz, F. Mormann, C. Rieke, P. David, C.E. Elger, Indications of nonlinear deterministic and finite-dimensional structures in time series of brain electrical activity: dependence on recording region and brain state, Physical Review E 64 (6) (2001) 1–8.

[3] A.L. Goldberger, L.A. Amaral, L. Glass, J.M. Hausdorff, P.Ch.Ivanov, R.G. Mark, J.E. Mietus, G.B. Moody, C-K. Peng, H.E. Stanley, PhysioBank, PhysioToolkit, and PhysioNet: components of a new research resource for complex physiologic signals, Circulation 101 (23) (2000) e215–e220.

[4] M. Golmohammadi, V. Shah, S. Lopez, S. Ziyabari, S. Yang, J. Camaratta, J. Obeid, J. Picone, The TUH EEG seizure corpus, in: American Clinical Neurophysiology Society, Phoenix, Arizona, USA, 2017, [Online] Available https://www.isip.piconepress.com/projects/tuh_eeg/. (Accessed 7 November 2017).

[5] Kai Fu, Jianfeng Qu, Yi Chai, Yong Dong, Classification of seizure based on the time-frequency image of EEG signals using HHT and SVM, Journal of Biomedical Signal Processing and Control 13 (2014) 15–22.

[6] Ram Bilas Pachori, Shivnarayan Patidar, Epileptic seizure classification in EEG signals using second-order difference plot of intrinsic mode functions, Journal of Computer Methods and Programs in Biomedicine 113 (2) (2014) 494–502.

[7] Andrea Varsavsky, Iven Mareels, Mark Cook, Epileptic Seizures and the EEG Measurement, Models, Detection and Prediction, CRC Press, 2011.

[8] 10-20 System Positioning Manual by Trans Cranial Technologies, 2012.

[9] M. Jalili, E. Barzegaran, M.G. Knyazeva, Synchronization of EEG: bivariate and multivariate measures, IEEE Transactions on Neural Systems and Rehabilitation Engineering 22 (2) (2014) 212–221.

[10] Jun Lu, Kan Xie, D.J. McFarland, Adaptive spatio-temporal filtering for movement related potentials in EEG-based brain–computer interfaces, IEEE Transactions on Neural Systems and Rehabilitation Engineering 22 (4) (2014) 847–857.

[11] V.K. Harpale, V.K. Bairagi, Significance of independent component analysis (ICA) for epileptic seizure detection using EEG signals, in: Proceedings of the International Conference on Data Engineering and Communication Technology (ICCUBEA2017), Pune, in Advances in Intelligent Systems and Computing, vol. 2, 2017, pp. 829–838.

[12] M.M. Hartmann, K. Schindler, T.A. Gebbink, G. Gritsch, T. Kluge, PureEEG: automatic EEG artifact removal for epilepsy monitoring, Neurophysiologie Clinique/Clinical Neurophysiology 44 (5) (2014) 479–490.

[13] N. Mammone, F. La Foresta, F.C. Morabito, Automatic artifact rejection from multichannel scalp EEG by wavelet ICA, IEEE Journal of Sensors 12 (3) (2012) 533–542.
[14] A. Hyvarinen, E. Oja, Independent component analysis: algorithms and applications, Pubmed Journal of Neural Network 13 (4–5) (2000) 411–430.
[15] M.Z. Parvez, M. Paul, Features extraction and classification for ictal and interictal EEG signals using EMD and DCT, in: Proceeding of the 15th International Conference on Computer and Information Technology (ICCIT), 2012, pp. 132–137.

CHAPTER 3

Seizure detection methods and analysis

3.1 EEG feature extraction

The data analysis of electroencephalography (EEG) remains challenging because the volume-conducted EEG suffers from a low spatial resolution. Multichannel EEG simultaneously measures coordinated brain activities at multiple sites on the scalp at millisecond temporal resolution, which makes it valuable for cognitive and neural engineering studies and clinical applications.

Quantitative analysis of EEG includes measurement of the mean spectral magnitude or power for multiple frequency bands. Quantitative analysis is the process of measuring specific properties of raw EEG signals and transforming them into digitally recorded EEG signals. Numerical parameters are then analyzed for various applications such as brain disorder detection and emotion analysis.

Electroencephalogram (EEG) is a tool which records a multifrequency nonstationary brain signal from various localized channels. Feature extraction is a necessary aspect of the analysis and classification of EEG signals. With respect to this study, it is a matter of extracting multiple features to detect epileptic seizures and differentiating brain signals into seizure, nonseizure, and preseizure EEG. Every research in EEG demands different related features for analysis.

There are linear as well as nonlinear tools for data analysis. Linear tools can be applied to nonlinear signals with a stationary assumption which may cause some loss of information. Nonlinear tools are best suited for nonstationary and nonlinear signals. Features can be the time domain, frequency domain, or time and frequency domain (TFD). Extracted features such as amplitude, periodicity, and synchronicity are used for providing qualitative information for classification of epileptic EEG.

The TFD EEG features are one of the basic features used for EEG analysis [1]. Some of the important features are given below.

(i) Mean of the dataset:

Brain Seizure Detection and Classification Using EEG Signals
https://doi.org/10.1016/B978-0-32-391120-7.00008-6

The mean of the dataset is calculated as

$$Mean(\mu) = \left(\sum_{i=t}^{t+s} x(i)\right)/n, \tag{3.1}$$

where $i = (t, t+1, \ldots, t+s)$ and $t = 1, 1+s, 2+s, a\ldots, N+s$, with μ being the mean of the signal, "x" the value of data at particular instant "i," "n" the number of data samples in the specified window size of "s," and "t" the first instant of the window sample.

(ii) Coefficient of variation:
The coefficient of variation (COV) is a statistical measure used to calculate the relative dispersion of data samples in a selected dataset per window sample at a particular instant around the mean of the samples:

$$COV_i = \delta_i/\mu_i, \tag{3.2}$$

where "δ" is the standard deviation observed in the signal and calculated as

$$\delta_i = \sqrt{\left(\sum_{i=t}^{t+s}(x_i - \mu_i)^2\right)/n - 1}. \tag{3.3}$$

(iii) Root mean square:
The root mean square (RMS) value of a signal x(t) is calculated as the square root of the average of the squared value of the signal, mathematically represented as

$$RMS_i = 1/T\left(\sqrt{\int_0^T x(t)^2 dx}\right), \tag{3.4}$$

where $T = t * s$, where 's' is number of samples due to windowing.

(iv) Kurtosis:
Kurtosis is a measure of the distribution of a signal, with higher values indicating a higher, sharper peak and lower values indicating a lower, less distinct peak. It is the fourth-moment estimation of a signal and given as

$$K_i = E(x_i - \mu_i)^3/\delta_i^4. \tag{3.5}$$

(v) Skewness:
Skewness is a measure of the asymmetry of the probability distribution. As per Bulmer (1969), if the skewness lies between −0.5 and

+0.5, then the distribution is symmetric; otherwise it is asymmetric. It is calculated as follows:

$$SKi = \sqrt{\sum_{i=1}^{n} (x - xi)^3 / (n - 1)\delta^3}. \tag{3.6}$$

(vi) Power spectral density:
The average of the Fourier transform magnitude squared over a large time interval is called power spectral density (PSD):

$$PSD_i = \lim_{x \to \infty} E\left(1/2T\left(\int_{-T}^{T} x(t)e^{-j2\pi ft}dx\right)\right). \tag{3.7}$$

The following spectral features are extracted to improve the overall system performance:

(a) The spectral centroid (SC): It is a measure of the amplitude at the center of the spectrum of the signal distribution over a window calculated from the Fourier transform frequency and amplitude information. It is the ratio of the average frequency weighted by amplitude to the mean of the signal:

$$SC_i = \sum_{k=0}^{n} f(k) * x(k)/m. \tag{3.8}$$

(b) The spectral slope (SS): It is a high-frequency response of spectrum calculated using linear regression and the central wavelet of the signal for a window. It is observed to be a useful feature in speech processing [28] and given by the following equation:

$$SS_i = (RE1 - RE0)/(\lambda 1 - \lambda 0), \tag{3.9}$$

where RE1 and RE0 are radiating energy with central wavelengths $\lambda 1$ and $\lambda 0$.

(c) The spectral flux (SF): It is a deviation of the power spectral response calculated from consecutive windows' power spectrum values. These variations are also called the temporal variation of the signal and usually used for onset detection of abnormal behavior. The spectral flux is calculated as follows:

$$SF = 1/N\sqrt{(P_i - P_{i-1})^2}. \tag{3.10}$$

(d) The spectral entropy (SE): It is a normalized probability density function (PDF) estimated from the signal and also treated as a measure of complexity[23]:

$$SE = -\sum PDF_i \, lnPDF_i. \tag{3.11}$$

These features are extracted using the TFD method, discrete wavelet transform (DWT), empirical mode decomposition (EMD), and singular spectrum EMD. The relevance of a feature is checked using analysis of variance (ANOVA) as a feature selection method, and suitable features are submitted to the classifier for seizure detection.

3.1.1 Time domain features

For the classification of epileptic EEG, time domain features used are an increase in amplitude, increase in regularity, and increase in synchronicity [2]. Average mean, variance, and variability estimates in EEG signals change over time with a suitable window length. Good temporal resolution and short events in the signal can be analyzed by short window size, and long-time behavior can be estimated by a long window. Repeatability and regularity (periodicity) of the signal are analyzed by the autocorrelation function. The autocorrelation for the nonstationary signal is not robust because of the largest changes during the seizure, i.e., during seizures the regularity of the signal increases and the EEG signal exhibits more oscillatory behavior. But autocorrelation can be part of the feature set to detect a seizure.

A particular signal is analyzed over a specific range of measurements, and its features are characterized by time, which is known as time domain analysis. The time domain analysis also estimates the synchronization of various EEG signals measured from various electrodes. A measure of synchronicity gives an idea of how similar signals are to each other. Synchronicity calculated explicitly with the linear cross-correlation of two different EEG signals can also be calculated by phase locking. Thus time domain features such as energy, power, mean, and variability show amplitude variation, the regularity is tested with variance, and the COV, total variation, and similarly synchronicity are verified with cross-correlation and phase locking.

For initial experimentation, a single-channel database [3] with a recording of tonic-clonic seizure from the C4 electrode and sampled with 102.4 samples/sec and a total duration of 3 minutes was considered.

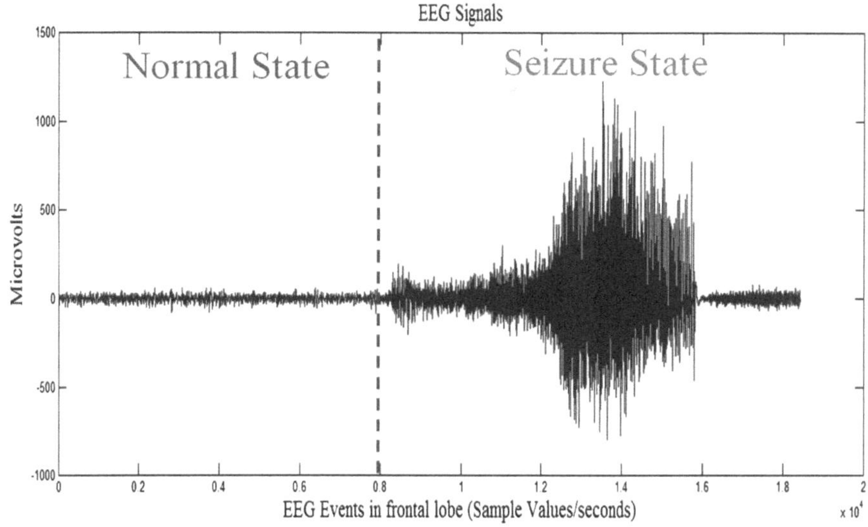

Figure 3.1 Reconstructed EEG signal.

Fig. 3.1 shows EEG signals showing a tonic-clonic seizure extracted from an electrode placed at the C4 location as specified by the international 10-20 system. The total number of samples is 18,420. It shows very little amplitude variation in EEG signals during the normal state (about 20–50 μV) and very high variation during the seizure state (up to 1000 μV). The seizure EEG signal plotted in Fig. 3.1 is a tonic-clonic seizure signal. The frequency or zero crossing rate (ZCR) is also very high during the seizure state. The mean value, as one of the time domain features, gives an idea about the amplitude of the signal at a particular time, as shown in Fig. 3.2. The frame rate considered for feature extraction is 512 samples/seconds. The normalized mean value is calculated for the signal shown in Fig. 3.2, and it shows that the range of the mean value is high during the seizure state compared to the normal state. Variance and standard deviation features define how the signal spreads over time and its regularity by calculating its deviation from the mean value, as shown in Fig. 3.3. At 27th window samples, maximum standard deviation is observed, thus this value is used for threshold selection. Thus, some of the time domain features such as mean power and standard deviation can help in understanding the increase in amplitude and regularity.

A standard deviation normalized by its mean value to extract the significance of various mean values of the signal is termed COV (Fig. 3.4).

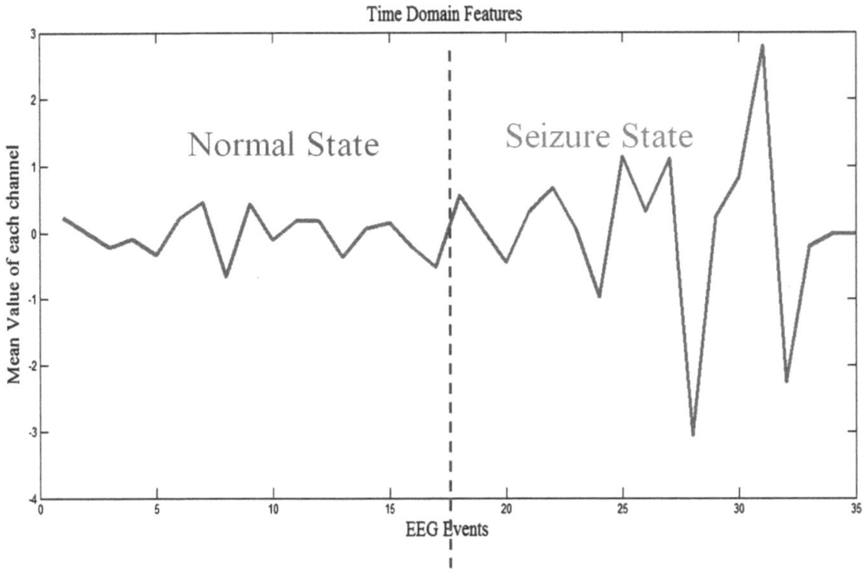

Figure 3.2 Mean value of single-channel EEG.

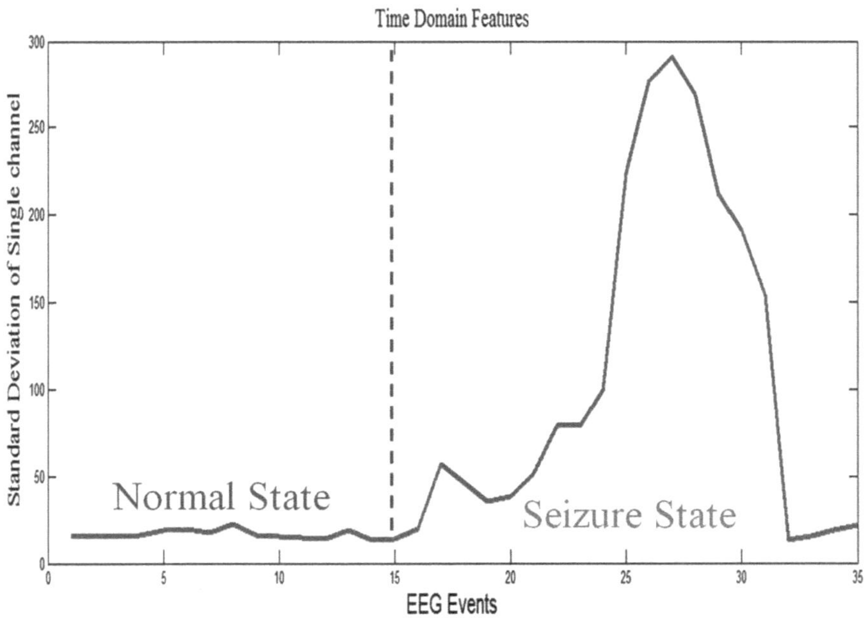

Figure 3.3 Standard deviation of single-channel EEG.

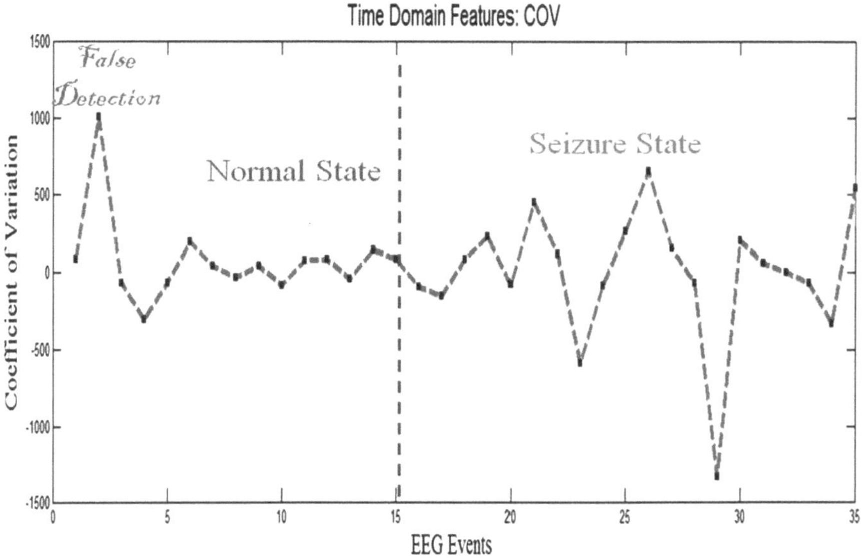

Figure 3.4 Coefficient of variation of EEG signal.

The COV feature is not significant as mean or standard deviation features. Variation in COV is higher during seizures but has the possibility of false positives, as observed at window sample 2.

Autocorrelation and cross-correlation help in finding synchronicity in various channels. Cross-correlation is useful for finding the correlation of different frequencies with the EEG signal. Fig. 3.5 specifies the correlation of brain activity with the EEG signal. It shows that the highest frequency component of about 35 Hz is observed during the seizure window. But this feature is not relevant as it does not exhibit a clear distinction between normal and seizure feature values.

These time domain features provide only spatial information, but temporal information is not addressed. During seizures, the signal shows very uncertain oscillations; thus, there is a need to separate frequency components to significantly classify the seizures. So along with time domain features, frequency domain analysis is also required to detect and classify all seizures.

3.1.2 Frequency domain features

Frequency is the measure of the occurrence of the events in a specified time. An EEG is a nonstationary signal comprised of events at different

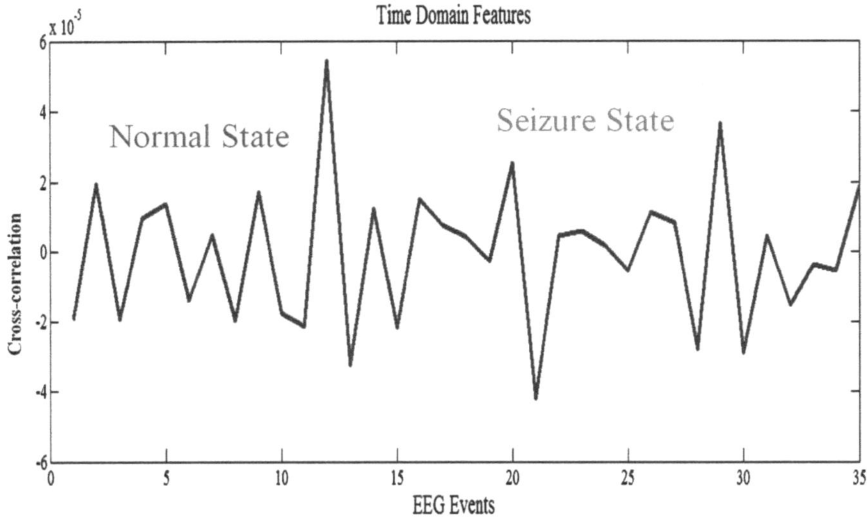

Figure 3.5 Time domain features extracted from the C4 channel.

frequencies. If a signal is represented by its frequency component and estimates all related features in frequency, this is known as frequency domain analysis [1].

In frequency domain analysis, the EEG signal is first transformed to the frequency domain from the time domain with Fourier transform or any other suitable transform. For discrete-time finite time domain signals the fast Fourier transform (FFT) of signal $x[n]$ for $n = k+1, k+2, \ldots, k+N$ is given by

$$FFT[\omega, k] = \sum_{n=1} Nx[n+k]e^{-j\omega n}, \tag{3.12}$$

where $m = 0, 1, 2, \ldots, N-1$. The value of the FFT at each ω represents the relative contribution of events that occur at that frequency to x[n]. To scale to the correct frequency range in Hz, ω is converted to $\omega = 2\pi f/Fs$, where Fs is the sampling rate of the data and f is the frequency in Hz between Fs/2 and Fs as per the Nyquist rate. The single-sided amplitude spectrum $Y(f)$ describes the frequency distribution over total samples of the EEG signal, as shown in Fig. 3.6. It shows most of the sample values are in the range of 0 to 45 Hz for tonic-clonic seizures recorded at the C4 electrode.

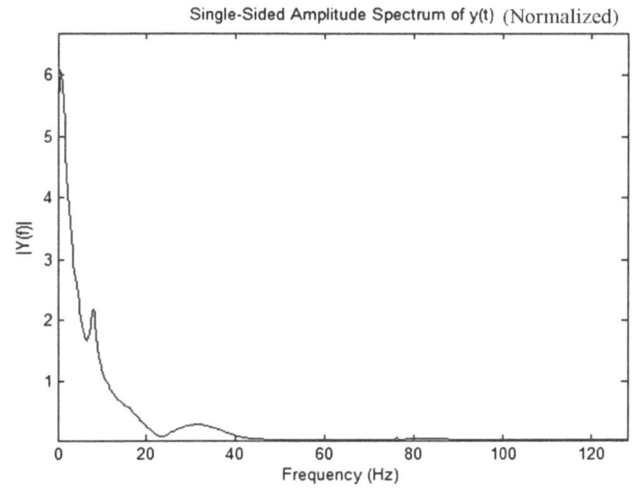

Figure 3.6 Amplitude spectrum of a single-channel EEG.

The power contributed to the overall signal $y[n]$ by a single frequency component is defined as the normalized PSD over ω, which is given as

$$PSD[\omega, k] = |FFT[\omega, k]|^2. \tag{3.13}$$

PSD analysis is a tool that helps to define the static and dynamic properties of the EEG; these static and dynamic properties describe the spatiotemporal behavior of EEG signals. Signal power for different events at their frequencies is calculated by PSD. Thus time domain and frequency domain representations contain the same information, but the features accentuated in each domain differ.

The PSD is estimated. It is the amount of power caused by events at each frequency. PSD has the limitation that it cannot isolate the time at which the frequency of interest occurs. In spectrography, the sequential computation of PSD over the small window is calculated to extract the frequency and temporal information for the detection of seizures in the EEG signal.

PSD is an important tool to understand local stationary behavior and capture the time-evolving nature of EEGs. This gives static and dynamic properties of the EEG signal. Fig. 3.7 specifies how much power of each frequency component contributes to the overall EEG signal of a single channel. For the detection of seizures, PSD alone is not sufficient; some

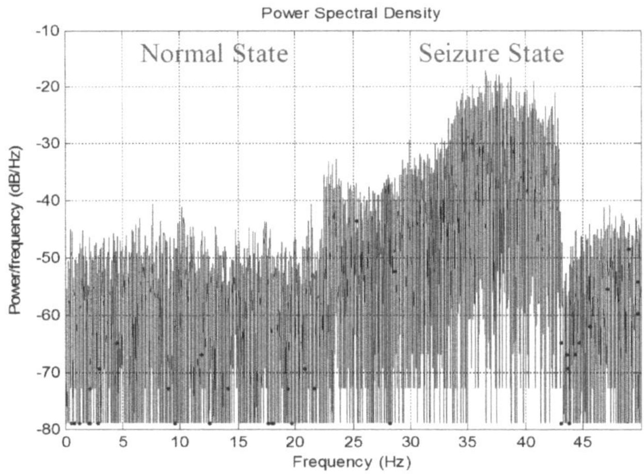

Figure 3.7 Power spectral density function of EEG.

features such as synchronization, periodicity, and energy of the signal must also need be analyzed.

3.1.3 Time-frequency analysis of EEG signals

Time domain analysis can provide better spatial information, but is poor in frequency content information required for EEG classification. The frequency domain can provide temporal information but only after windowing the function. To get optimal results, the window size should be perfectly estimated. Selection of window size is the biggest challenge in frequency analysis. Time-frequency analysis resolves both of these problems. The data analysis methods studied in this section are considered as linear systems, but EEG signals are nonlinear. To analyze signals with less computation, they are considered as linear signals over a small window (linear approximation). Various methods have been proposed to extract information from EEGs with efficient performance measures.

1. **Time-frequency analysis with Fourier transform**
   The most commonly used method of frequency analysis to estimate the frequency component in the EEG signal was the Fourier transform during the early 1990s. Because of its high computational speed, it was preferred in real-time monitoring and analysis of various physiological signals. Fourier transform allows separation of various EEG rhythms, which facilitates analysis of the occurrence of rhythmic activities in signals. FFT analysis is applied on specific time intervals of EEG data, with

each time interval composed of pre-event and post-event stimuli. In FFT analysis, these time intervals are predefined. Nonstationary EEG signals contain artifacts, but in FFT analysis artifacts-free signal data are preferable. Before computing the Fourier transforms, each epoch is multiplied by a proper windowing function; preferably a Hanning window is used, which handles border problems. These periodograms are computed and averaged for further analysis. Parameters that can be observed with the Fourier transform are relative power (power ratio of alpha activity and theta activity), reactivity (ratio of alpha activity during sleep and nonsleep states of the brain), and the asymmetric index. As EEG is a multifrequency signal, Fourier transform is useful in EEG analysis with limitations such as its stationarity, but EEGs are highly nonstationary and time progression of the frequency patterns is lost.

2. **Time-frequency analysis with Gabor transform (short-time Fourier transform)**
   The Fourier transform is a frequency-based function derived from complex-valued parameters and gives information about the amount of the frequency component present in time domain signals, but there is a lack of information about the time evolution of the frequencies. As EEG has time-varying features, it cannot be resolved with Fourier transform. Fourier analysis cannot detect real-time frequency variations. This problem can be partially resolved by Gabor transform, also called as Short-Time Fourier transform (STFT). It is a windowed Fourier Transform in which the Fourier Transform is progressively taken over a time window of a few seconds with stationary window length. Thus the nonstationary signal is divided in time segments, and the Fourier transform is successively applied to each segment. EEG analysis with Gabor transform facilitates the identification of tonic-clonic seizures and provides quantitative measures of the dynamics of epileptic seizures. Gabor transform has the limitation that its window length is predefined and cannot vary as per the requirements, so window selection is a challenge in Gabor transform. Narrow windows give a poor frequency resolution, and wide windows render the time localization nonprecise. This causes uncertainty in data analysis.
3. **Time-frequency analysis with wavelet transform**
   Limitations of the STFT related to the selection of window size can be overcome by wavelet transform. The wavelet transform has the capability of varying frequency based on the frequency component (narrow windows for high frequencies, broad windows for low frequencies). It

facilitates optimal time-frequency resolution in all frequencies, and it eliminates the requirement of signal stationarity. Recent research suggests wavelet analysis is the best method for time-frequency analysis.

A wavelet $\psi[n]$ is a function defined as

$$\sum_{-\infty}^{\infty} \psi[n] = 0. \tag{3.14}$$

Power is defined as

$$\sum_{-\infty}^{\infty} (\psi[n])^2 < \infty. \tag{3.15}$$

A wavelet function shifted by "y" samples and scaled by "x" is given by

$$\psi_{xy}[n] = \frac{1}{\sqrt{x}} \psi\left[\frac{n-y}{x}\right]. \tag{3.16}$$

The mother wavelet is observed with $x = 1$ and $y = 0$. If $0 < x < 1$ contracts in time and when $x > 1$, $\psi xy[n]$ stretches in time. For large values of "x" the wavelet function is long. With different scales, the "x" temporal and frequency content can be extracted with different resolutions.

The continuous wavelet transform (CWT) is the correlation between the signal and the wavelet function. Calculating wavelet coefficients with CWT for every possible scale requires much effort and results in huge data. Therefore DWT is a preferable choice for EEG signals. The DWT is a filtering method to choose subsets of the scales (time and frequency) and positions of the wavelet mother function.

Fig. 3.8 shows generalized characteristics during Fourier transform, STFT, and DWT analysis. Thasneem Fathima et al. [4] used the wavelet transform for seizure detection using EEG analysis. The feature vectors are calculated on the basis of approximated wavelet coefficients of the EEG signals. DWT decomposes the signal into time-frequency representations to depict their distribution. This time-frequency distribution of the EEG signals can be characterized by minima and maxima of the wavelet coefficients, mean, and standard deviation of the wavelet coefficients in each subband.

Robert Jenke et al. [5] used wavelet transforms for emotion analysis and compared various algorithms for emotions by spectral analysis of EEG signals. Hemant K. Sawant et al. [6] proposed a method which utilizes discrete Fourier transform and the DWT for experimentation. Mainly, the wavelet

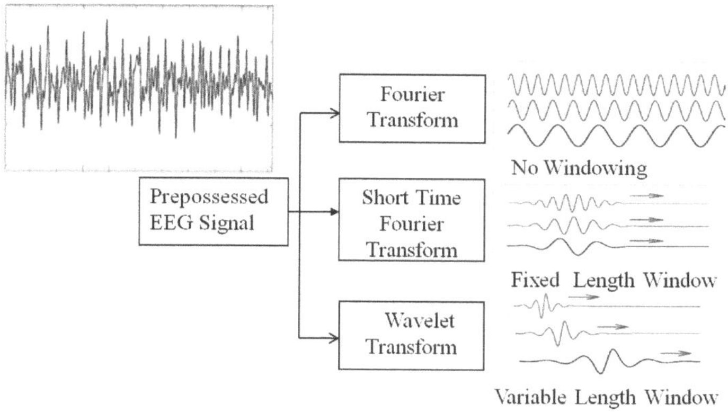

Figure 3.8 EEG signal analysis with Fourier transform, STFT, and discrete wavelet transform.

transform is used as a classifier of the EEG frequencies. Thus this method is a highly reliable and accurate multistage technique for recognition and classification of EEG waves.

Mariel Rosenblatt et al. [7] compared DWT and ODWT for quantitative analysis of EEG signals for tonic-clonic epileptic seizures. This method covers parameters such as wavelet statistical complexity and the normalized Shannon wavelet entropy defining the energy of different frequency bands.

S. Nasehi et al. [8] present a simplified Gabor wavelet transform (GWT) for emotion recognition and analysis where spectral, spatial, and temporal features were extracted with GWT and classified with a neural network. This method achieved a reduction in computation cost of emotion analysis. Maeda et al. [9] proposed and implemented a modified GWT with sigmoid function responsible for adjusting the damping coefficient of Gabor wavelets. This method achieved adequate time-frequency resolution for sleep EEG. It makes differentiation of Fourier transform-, STFT-, and wavelet transform-based EEG analysis possible. There are various advanced transform techniques such as dual-tree complex wavelet transform and stationary wavelet transform.

Thus various features such as time domain, frequency domain, and TFD features are extracted for seizure detection. These features are useful in identifying random oscillations, periodicity, and synchronicity of the EEG signal. This helps detecting seizures in EEG signals.

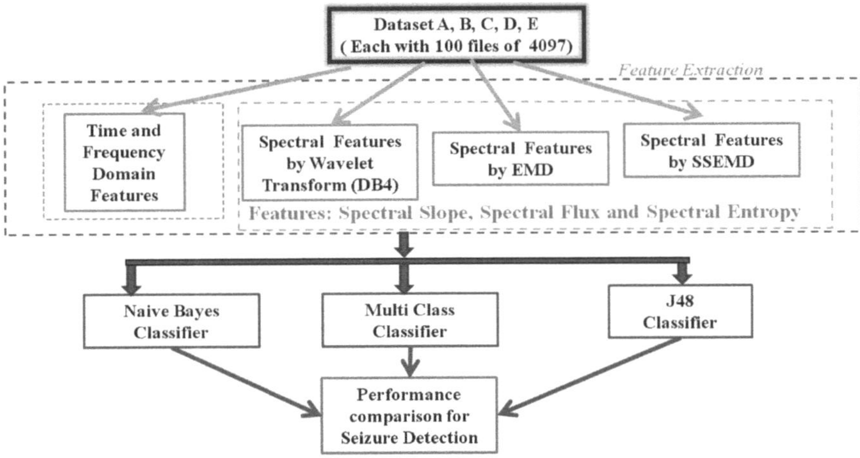

Figure 3.9 Implementation flow of seizure detection.

3.2 Improved seizure detection methods

The outline of the proposed methodology is explained in Fig. 3.9. TFD features such as mean, RMS, the COV, total variation, standard deviation, kurtosis, PSD, and entropy are extracted. Similar features along with spectral features are extracted from wavelet coefficients and EMD. These features are analyzed, and an effective feature vector is constructed. These feature vectors are applied to various high-performance machine learning classifiers to analyze the performance of the system.

3.2.1 Dataset

The standard EEG database from Andrzejak (2001), described earlier, is used for experimentation and evaluating the proposed methodology. The dataset is comprised of five sets (A, B, C, D, and E) with 100 EEG segments with a sampling frequency of 173.60 Hz and a recording duration of 23.6 seconds. In the proposed experimentation, set A was recorded from normal subjects with eyes open, set B was recorded from normal subjects with eyes open, and set E was recorded from epileptic patients during seizures. The signals are classified into three classes, viz. Normal_o, Normal_c, and Seizure. In total, 300 files with 4097 samples were used for experimentation.

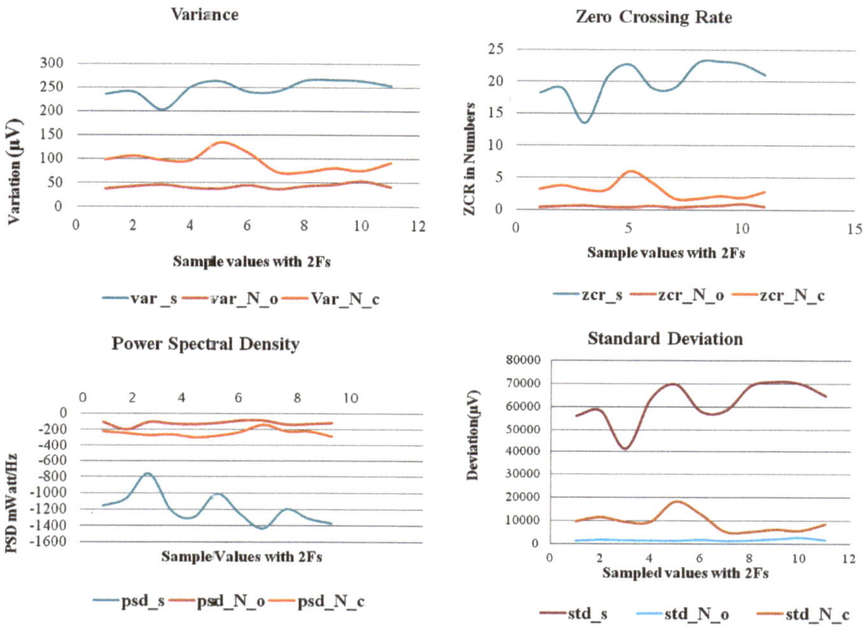

Figure 3.10 Time and frequency domain features extracted for Normal_o, Normal_c, and Seizure.

3.3 Feature extraction for seizure detection

The features extracted based on time and frequency are mean value, RMS, standard deviation, COV, power, ZCR, PSD, minima, and maxima, as defined earlier. Some of the features are shown in Fig. 3.10. These features are calculated for the normal signal with eyes open (Normal_o), the normal signal with eyes closed (Normal_c), and the seizure onset state. The variation of the signal is visualized, but not distinctly modeled for nonstationary conditions. Detection of seizures can be a challenging problem due to myogenic artifacts, interference, and the combined impact of the mental state of the patient. Thus, seizures need to be modeled with appropriate methods.

3.3.1 Time and frequency domain features

For classification of epileptic EEGs, time domain features used are an increase in amplitude, increase in regularity, and increase in synchronicity. [2]. Average mean, variance, and variability estimates change in EEG signals over time with a suitable window length. A good temporal resolution

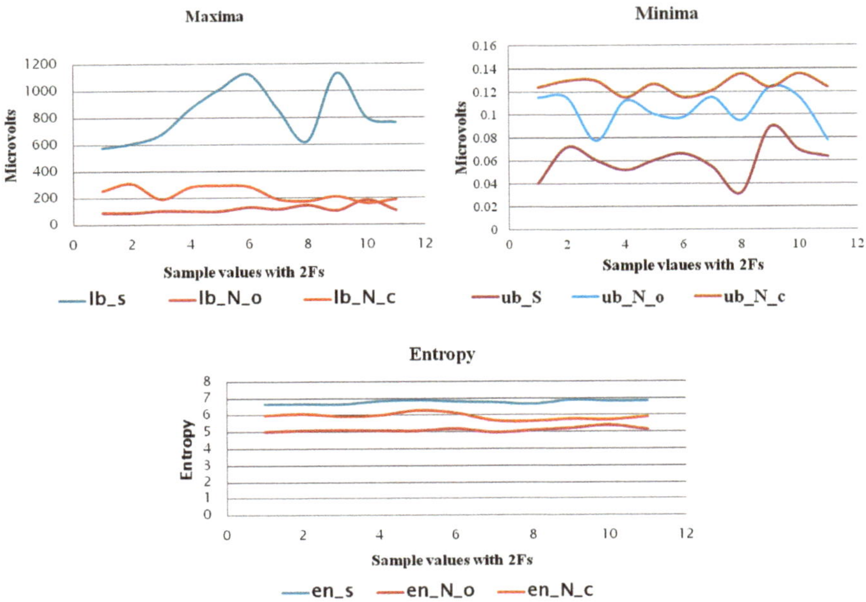

Figure 3.11 Time and frequency domain features extracted for Normal_o, Normal_c, and Seizure continued.

and short events in the signal can be analyzed using short windows, and long-time behavior can be estimated using long windows. Repeatability and regularity (periodicity) of the signal are well analyzed by the autocorrelation function. Frequency is the measure of the occurrence of the events in a specified time. If a signal is represented by its frequency component and estimates all related features in frequency, this is known as frequency domain analysis. The features extracted on the basis of time are variance (VAR), ZCR, PSD, and standard deviation (Fig. 3.10). It shows that feature values observed during seizure and normal states are different, thus these are relevant features. Fig. 3.11 shows time base variation with maximum and minimum amplitude variation along with entropy. These features help in deciding the thresholds required for classification. Similarly, Fig. 3.12 shows the mean and RMS values with 2Fs window. It shows mean feature values do not clearly distinguish between normal and seizure values, so these cannot be used for further classification, whereas RMS is a relevant feature. The most commonly used frequency domain feature is PSD. The normalized PSD estimates the amount of power caused by events at each frequency. PSD has the limitation that it cannot isolate the time at which

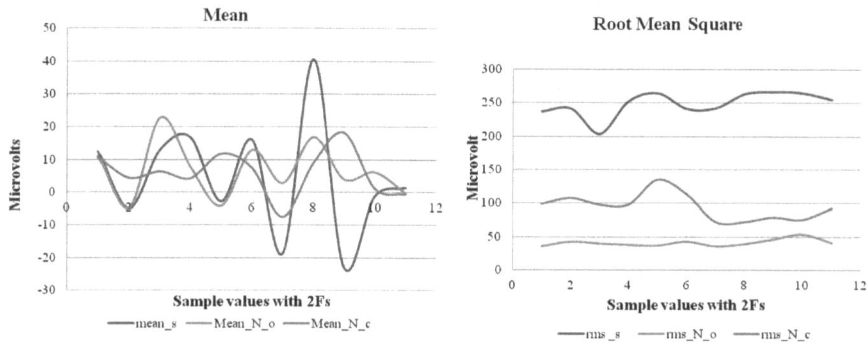

Figure 3.12 Time domain features extracted for Normal_o, Normal_c, and Seizure continued.

frequencies of interest occur. Spectral entropy denotes the uncertainty of the data and defines complexity. In Figs. 3.10–3.12, each of the plot, except mean value, shows a clear distinction in values of parameters between Normal_o, Normal_c, and Seizure. For example, ZCRs plotted for zcr_s, zcr_N_o, and zcr_N_c show separable characteristics for the three classes.

3.3.2 Feature extraction using discrete wavelet transform

DWT is a useful analysis method for multifrequency EEG signals compared to CWT [4]. The method adopted in this research includes EEG signal decomposition using the Daubechies wavelet function (DB4) with four levels of decomposition or five subbands called delta (0.1 to 3.5 Hz), theta (4 to 7.5 Hz), alpha (8 to 13 Hz), beta (14 to 30 Hz), and gamma (>30 Hz) bands, as shown in Fig. 3.13, which shows DWT decomposition of a normal EEG signal with its energy distribution. The Daubechies wavelet function is available with orders of 2, 4, 6, and 8. All of the functions are well proven because of its orthogonal property and effective filter operation. But DB4 is observed to be an appropriate filter function for epileptic seizure detection [10]. As small-order functions could not distinctly detect peaks of the EEG signal and higher-order functions have high oscillations, DB4 is selected for further analysis. The EEG signal is decomposed to level 4, and Detail coefficients D1, D2, D3, and D4 and Approximation coefficient A5 are decomposed brain waves. Fig. 3.14 shows decomposed brain waves for the normal state of an epileptic patient and Fig. 3.15 shows wavelet decomposition in five bands of the seizure signal. DWT filter decomposition gives separate frequency bands corresponding to various brain activities; for example, delta waves are associated with deep sleep, theta waves with normal

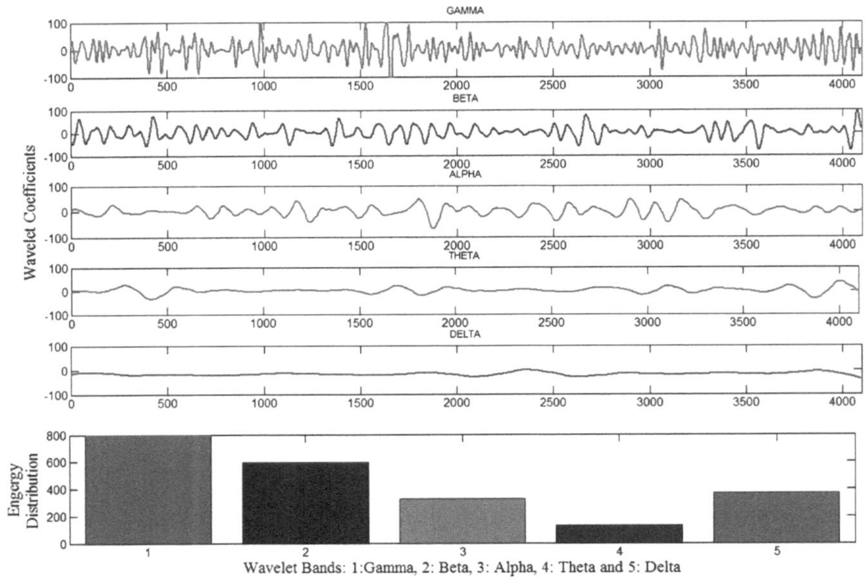

Figure 3.13 Wavelet decomposition in five subbands for EEG signals from normal patients.

sleep, alpha waves with the resting state, beta waves with the alert and active state, and gamma waves with the fastest state of the brain.

The standard features are extracted from each wavelet coefficient, such as variance, power, ZCR, and PSD, for all the frequency bands. As per the significance of seizure and normal features, they are statistically checked with ANOVA; it is observed that gamma and beta bands are more relevant for seizure detection.

The normalized features for gamma and beta bands are shown in Fig. 3.16 and Fig. 3.17, showing distinct characteristics during normal and seizure signals, thus indicating they are useful for classification. The spectral features are also extracted from gamma and beta frequency bands to improve the classification accuracy. These features are spectral slope, flux, centroid, roll-off, crest, entropy, and tone power ratio (TPR). Some of the relevant spectral features from the gamma and beta frequency bands are used to construct a further feature vector for classification, as shown in Fig. 3.18. These spectral features show differentiation in values during seizures and are thus significant for classification. All the features are normalized and compared for gamma and beta bands, as shown in Fig. 3.18. The spectral features are

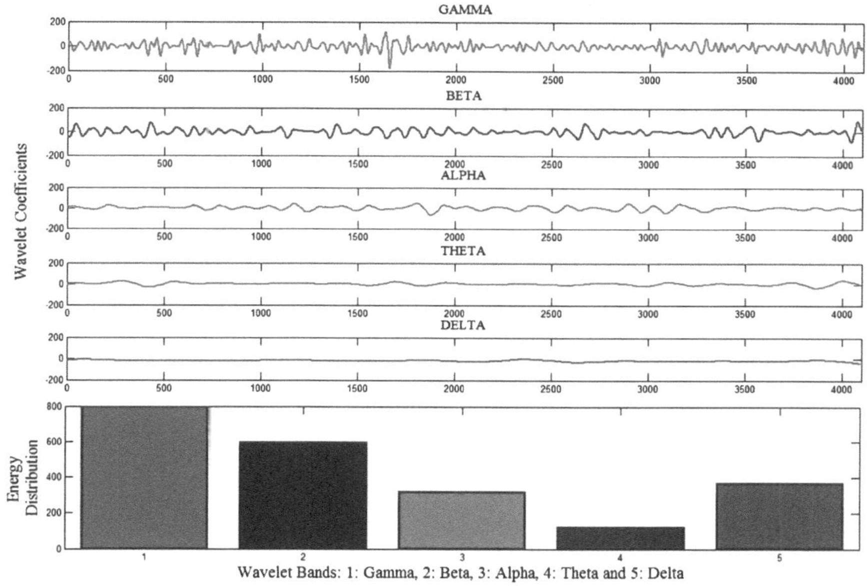

Figure 3.14 Wavelet decomposition in five subbands from epileptic patients during the normal state.

observed to be more relevant due to variations in frequency and amplitude during seizure activity.

3.3.3 Features extracted using empirical mode decomposition

EMD is a preferred method for the analysis of nonstationary physiological signals. In comparison with wavelet decomposition, EMD is proven to be a more effective method for EEG signal analysis as wavelet decomposition is fully dependent on the selection of the basis function, whereas EMD has the heuristic approach of decomposition [11]. The recently developed method decomposes a signal into a sum of components, each with slowly varying amplitude and phase [12].

EMD is an adaptive method; that is, the decomposition it produces is specific to the signal being analyzed. In theory, the components it outputs should separate phenomena occurring on different time scales. EMD is an adaptable approach to extract the nonstationary part of the signal. In EMD, the signal is decomposed into amplitude-modulated/frequency-modulated signals called intrinsic mode functions (IMFs) [13] and observed (Fig. 3.19). The IMF is selected in such a way that an IMF may only have one zero be-

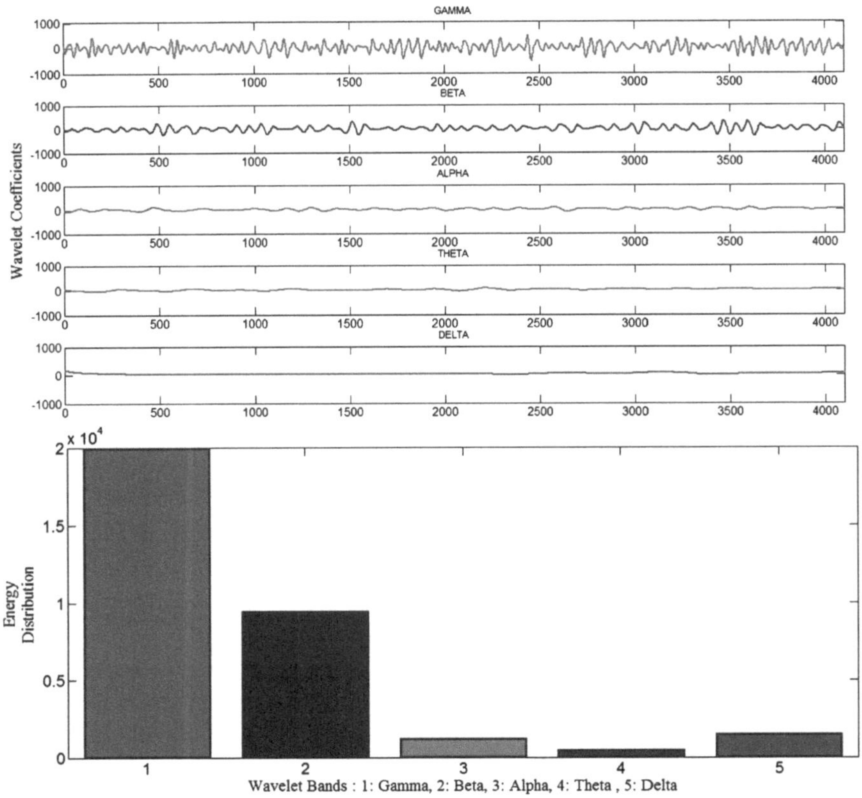

Figure 3.15 Wavelet decomposition in five subbands from epileptic patients during seizures.

tween successive extrema and an IMF must have zero local means. Belhadj et al. [14] introduced a novel method of epileptic seizure detection using a hybrid model of clustering and EMD.

This system uses EMD for seizure detection using time, frequency, and spectral features such as spectral entropy extracted from relevant IMFs to improve classification performance [15]. The feature vector is constructed for classification on the basis of its relevance with reference to seizure features. Fig. 3.20 and Fig. 3.21 show relevant IMF features such as entropy, PSD, and ZCR extracted from IMF1 and IMF2. The selection of IMF1 and IMF2 from five decomposition signals is discussed in section 3.3.5. Fig. 3.22 shows spectral features extracted from IMF1 and IMF2, showing their significance in normal and seizure signals. All features extracted from

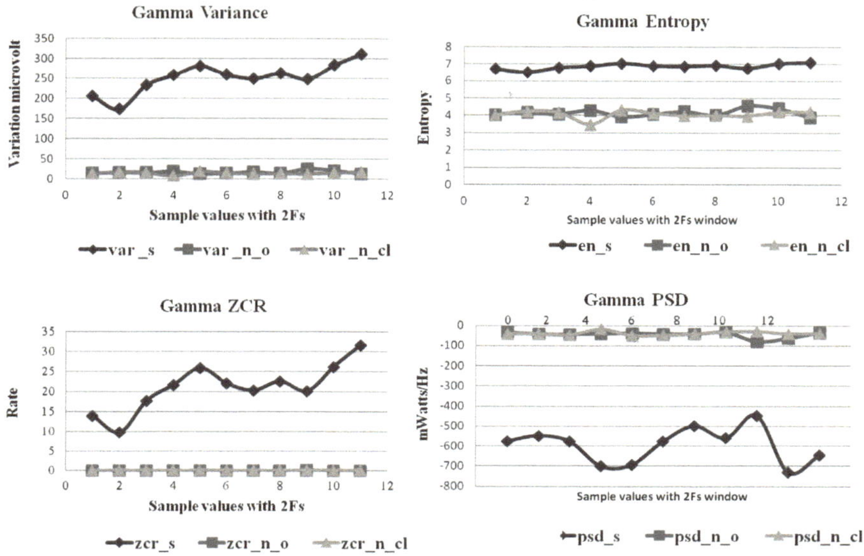

Figure 3.16 Significant time and frequency features extracted from the gamma band.

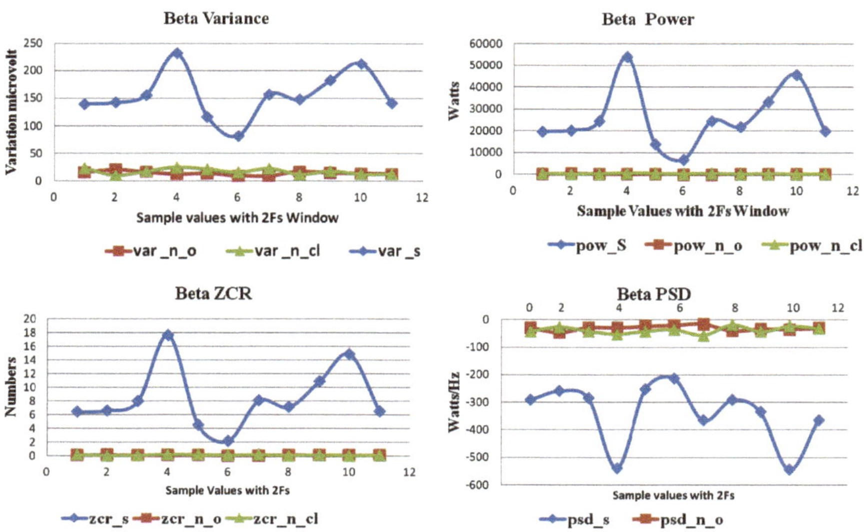

Figure 3.17 Significant time and frequency features extracted from the beta band.

files for normal patients recorded with eyes closed are referred to as Normal_c, the file for normal patients recorded with eyes open is referred to

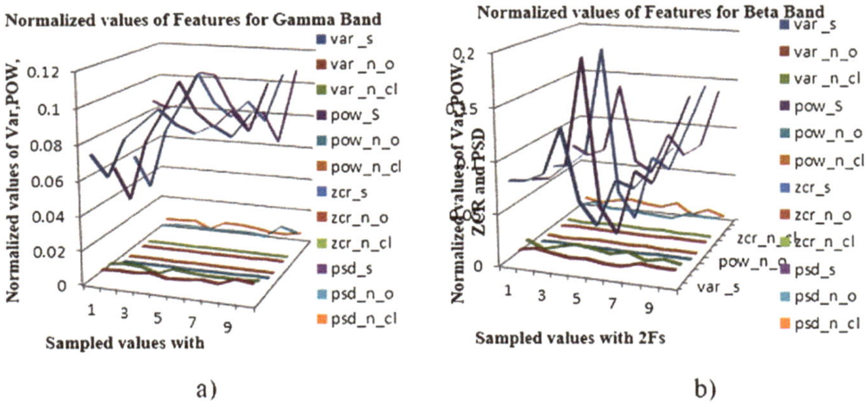

Figure 3.18 Normalized TFD features extracted from (a) the gamma subband and (b) the beta subband.

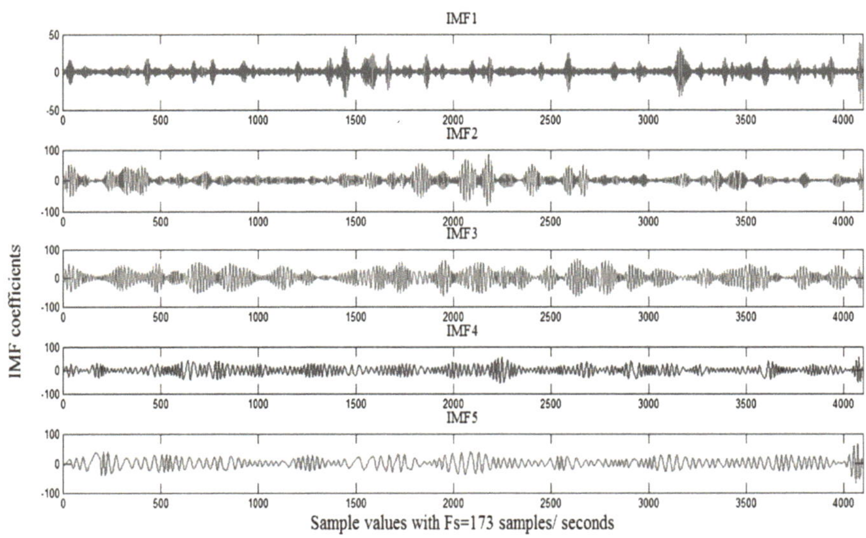

Figure 3.19 EEG signal decomposed in five IMFs.

as Normal_o, and the file recorded from seizure patients are referred to as Seizure. The total number of window samples plotted is 22: 11 samples of IMF1 and 11 samples of IMF2.

EMD-based features provide significant classification accuracy. In the next section, a new method to improve the false positive rate and accuracy named singular spectrum EMD (SSEMD) is presented.

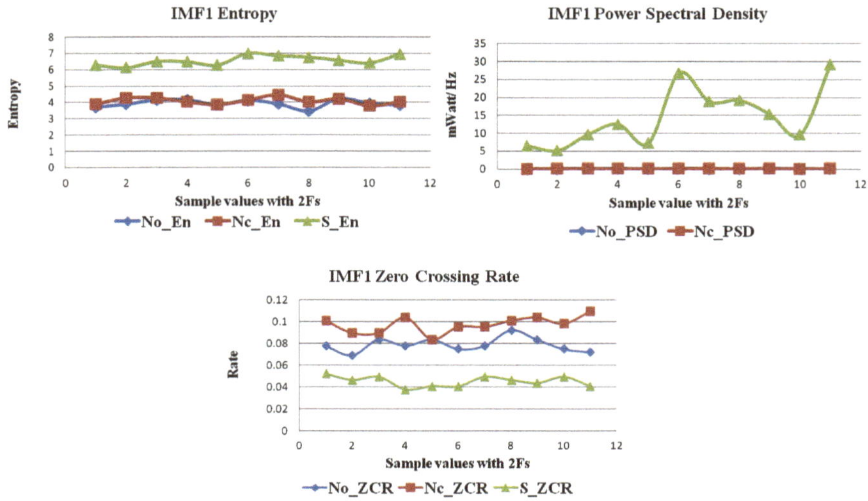

Figure 3.20 Time-frequency features from IMF1.

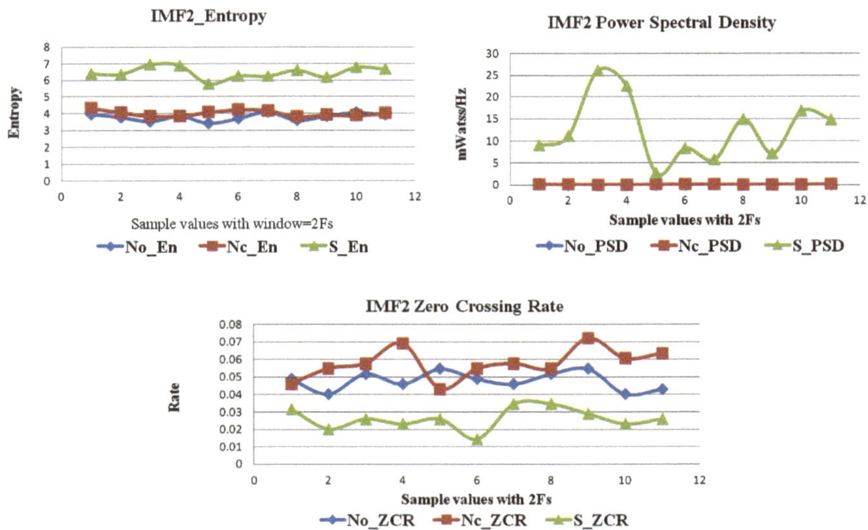

Figure 3.21 Time-frequency features from IMF2.

3.3.4 Feature extraction by singular spectrum empirical mode decomposition

EMD is different from the wavelet transform as it does not predict any behavior of the signal and thus does not use any basis function. The IMF is

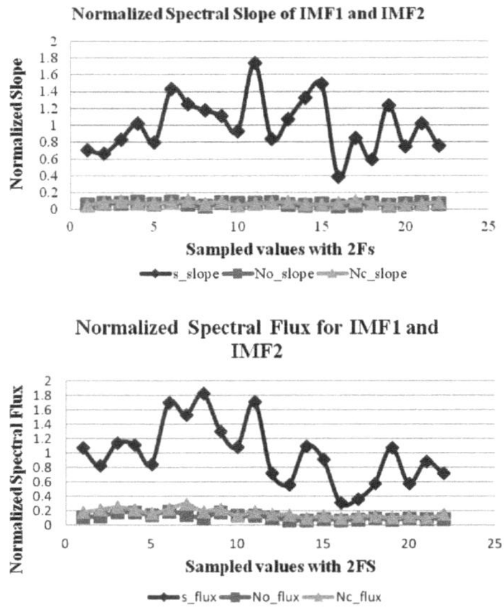

Figure 3.22 Spectral features extracted from IMF1 and IMF2.

derived by interpolation of maxima and minima of the signal. It does not have any predictive model for IMF, and also there is a lack of selection of the number of IMFs that could clearly describe the applied signal. The singular spectrum EMD (SSEMD) method is used to overcome the problem of refinement of IMF behavior [16]. This leads to decision of the narrow boundaries of the IMF and thus improves the classification accuracy of the seizure detection technique. This singular spectrum is an adaptive and nonparametric estimation method used for signal processing [17]. The spectral decomposition of the signal is observed, and suitable IMFs with narrow boundaries are constructed for effective feature extraction [18]. The steps involved in SSEMD [16] are as follows:

(i) Construct a covariance matrix either by the Toeplitz approach or by the trajectory approach.

(ii) Construct principal components by calculating eigenvalues and vectors.

(iii) Reconstruct the original signal, removing unwanted information from data.

(iv) Construct fine IMFs by EMD for feature extraction of seizure activity.

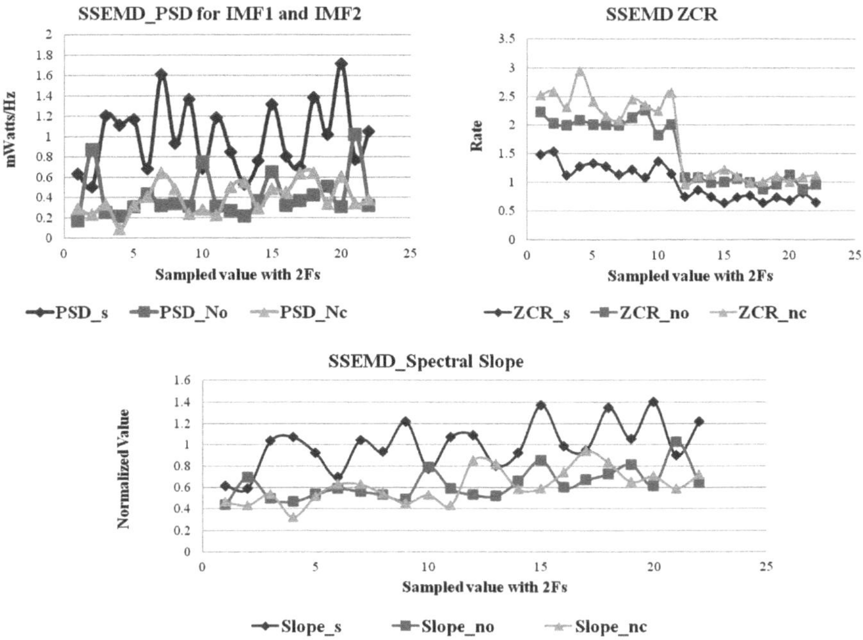

Figure 3.23 SSEMD-based features: power spectral density (PSD), zero crossing rate (ZCR), and spectral slope.

(v) Select appropriate IMFs and features for classification of the EEG signal into normal and seizure classes.

Fig. 3.23 shows the features extracted by the SSEMD method. It shows that feature values for Seizure are in a different range than features values for Normal_o and Normal_c. The normal values are in the same range; thus, the separable feature improves the classification accuracy.

3.3.5 Feature selection and feature vector construction

Nai-Fu Chang et al. [19] discussed the need for feature selection in EEG and the ECoG database using a machine learning approach. Appropriate feature selection or channel selection reduces the size of the feature vector pattern, avoids irrelevant computation, and improves classification or prediction performance. As the database used here is preprocessed with 173 samples/seconds and recorded for 23.6 seconds, it has a total dimension of 4096 samples for each of the subjects. When a rectangular window of 346 samples (2 × Fs) is used and 13 features are extracted from the data, the in-

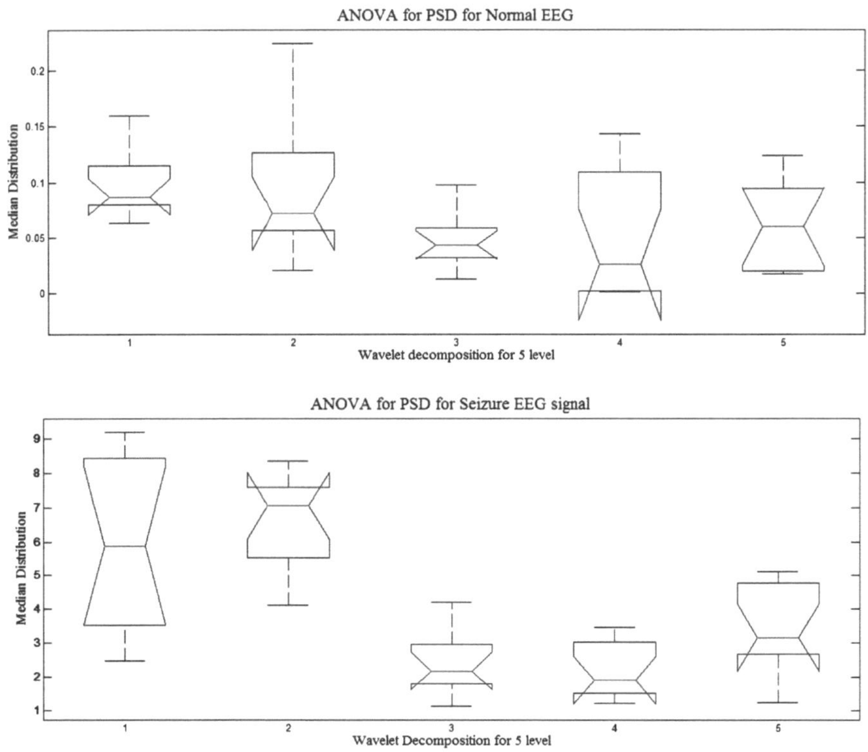

Figure 3.24 Boxplot for the median distribution of normal EEG and seizure EEG signal.

formation matrix will be of dimension (13 features) × (4096/346 window samples) × (number of subjects for training). Fig. 3.24 shows the median distribution of PSD features extracted from wavelet-filtered bands for normal EEG signals and seizure signals. It is clearly observed that for normal signals, the median distribution is in the range of 0 to 2, whereas for seizure signals the gamma band and the beta band have different distributions than the alpha, theta, and delta bands, which lie in the normal range of distribution. Feature extraction using wavelet transform, EMD, and empirical wavelet transform provides this feature matrix with five times higher dimensions than time and frequency features. These complex data lead to an increase in the complexity of the classification model and the required time of execution. This excess of matrix dimensions causes an overfitting problem. Thus the relevant or useful features need to be selected for improving classification performance.

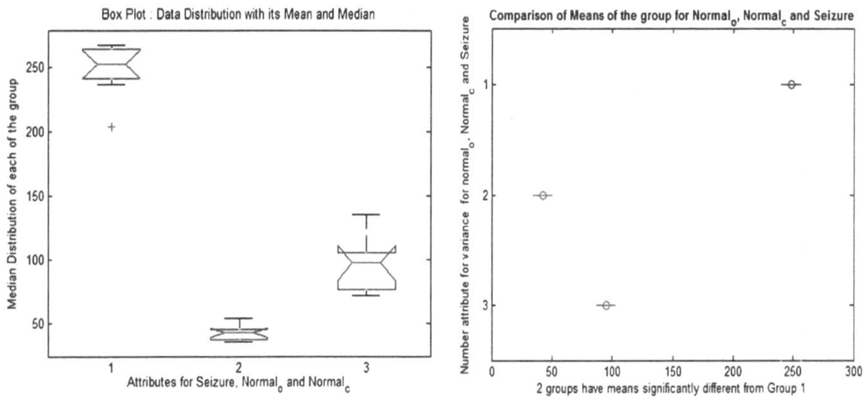

Figure 3.25 One-way ANOVA for PSD values. (a) Comparison of the mean distribution. (b) Boxplot showing the median distribution.

The "curse of dimensionality" problem addresses the reduction in the dimension of the feature set by combining the features, but the feature selection method either uses the corresponding feature or neglects the features on the basis of their relevance to the classification problem. The methods are used for feature selection, wrapping, and embedding. There are various algorithms for feature selection, such as the chi-square test and mRmR. One-way ANOVA is a well-established feature selection method and is used in various applications, such as e-mail spam classification [20], microarray data classification [21], and seizure detection [22]. This work uses one-way ANOVA for the selection of features, significant frequency bands, and IMFs.

Firstly, the EEG signals are preprocessed by removing artifacts and powerline noise [23] and then features are extracted as discussed in earlier sections in EEG preporocessing and feature extraction sections. One-way ANOVA is used to select relevant features from a set of TFD features. Features are distinctly different for the three groups Normal_c, Normal_o, and Seizure. In the presented work, 10 features are extracted, and only four features are selected for further experimentation, as discussed below.

Fig. 3.25(a) shows a boxplot representation of the median distribution using one-way ANOVA for the PSD feature to state its significance for classification. The boxplot shows the variation of the median distribution for seizures, whereas in the normal state, the shape of the boxplot shows that PSD values are distributed closer to the median and mostly overlaped. It also shows that the PSD distribution during the seizure state is significantly

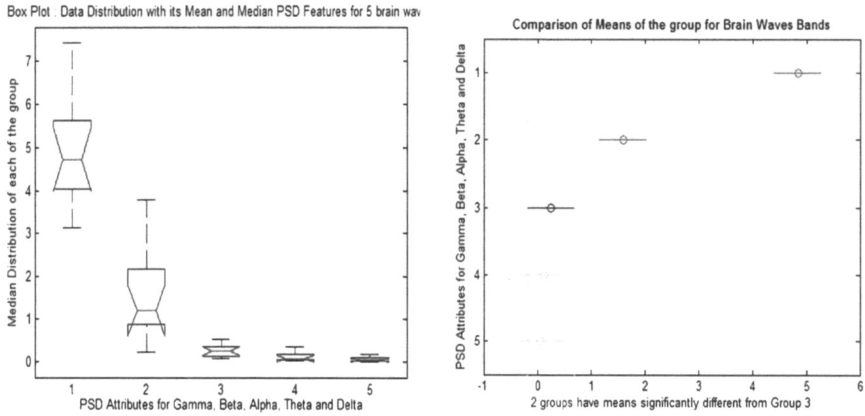

Figure 3.26 Boxplot showing a comparison of ZCR values of IMF1 to IMF5.

different from the PSD distribution in the normal state, based on $P < 0.05$ and the high value of the F-measure as compare to F-critical value.

The boxplot in Fig. 3.25(b) shows the data samples are distributed in four quartiles. Short quartiles indicate that the feature values are similar, and large quartiles indicate that the feature values differ more.

Using the P-value and the F-measure, relevant features are selected. Features used for classification using the TFD method are shown in Table 3.1. Features with $P < 0.05$ and very high F-measure values are selected for classification. Thus variance (Var), standard deviation (STD), RMS, power, ZCR, PSD, and entropy are observed to be relevant. Similarly, all the features were verified and a final feature vector was constructed for TFD, as shown by the following equation:

$$F1 = f(Var,\ PSD,\ ZCR,\ entropy). \tag{3.17}$$

For wavelet-based features, the signal is decomposed as gamma, beta, alpha, theta, and delta. The relevant variation in EEG signal in seizure patients is observed for these five bands using one-way ANOVA, as shown in Fig. 3.26(a) and (b), which shows a comparison of the median distributions over five bands. It is observed that gamma and beta have different means and median distributions than the remaining frequency bands during seizure onset. The various TFD features are calculated, and one-way ANOVA is applied to each of the features, taking seizure features as reference features. Similarly, ZCR has a significant average variation between

Table 3.1 ANOVA values for TFD features (# higher *P*-values and lower F-measures).

Para, feat	Mean	Var	STD	RMS	Power	COV	Kurt	ZCR	PSD	Ent
F-measure	0.09	503.18	384.3	508.05	377.63	0.14	0.10	377.63	290.24	366.59
P-value	0.90	8.39×10^{-24}	4.17×10^{-22}	7.29×10^{-24}	5.39×10^{-22}	0.86	0.90	5.38×10^{-22}	2.35×10^{-20}	8.26×10^{-22}

seizure and nonseizure activity. It is observed that out of the extracted features, PSD and ZCR are observed to be highly relevant. The feature vector constructed with wavelet transform is as follows:

$$F2 = f(PSD_{gamma},\ ZCR_{gamma},\ PSD_{beta},\ ZCR_{beta}). \tag{3.18}$$

The spectral features are extracted and analyzed using one-way ANOVA for identification of redundant and relevant spectral features. The relevant spectral features used for classification are spectral slope (SS), spectral flux (SF), and spectral entropy (SE). After considering TFD and spectral features, the feature vector F2 is modified to the hybrid feature model given by

$$F3 = f(PSD_{gamma},\ ZCR_{gamma},\ SS_{gamma},\ SF_{gamma},\ SE_{gamma}), \tag{3.19}$$

$$F3cont = (PSD_{beta},\ ZCR_{beta},\ SS_{beta}, SF_{beta},\ SE_{beta}). \tag{3.20}$$

The selected feature vector F3 improves the classification accuracy and the false positive rate. For EMD-based features, it is observed that the effective IMFs are IMF1 and IMF2 as per one-way ANOVA of all five IMFs with reference to ZCR features of seizure files as described by the boxplot shown in Fig. 3.27. The comparison of the mean shows that the average ZCR values of IMF1 and IMF2 have distinct relevance and can be used for classification. Thus features are extracted only for IMF1 and IMF2. Extracted features are again tested using one-way ANOVA to identify the most relevant features for classification. Similarly, the entropy feature for IMF1 and IMF2 shows a significant difference in entropy between the Seizure, Normal_o, and Normal_c groups. It is concluded that entropy, PSD, standard deviation, and ZCR are the most relevant features and can be used for classification. The feature vector constructed for EMD is given by

$$F4 = f(I1_{En},\ I1_{PSD},\ I1_{STD},\ I1_{ZCR},\ I2_{En},\ I2_{PSD},\ I2_{STD},\ I2_{ZCR}). \tag{3.21}$$

Furthermore, to improve the false positive rate and the accuracy of the system, spectral features are extracted for IMF1 and IMF2. It is observed that spectral slope and spectral flux carry the most useful information. Thus, the feature vector changes to F5, given by

$$F5 = f(I1_{En},\ I1_{PSD},\ I1_{STD},\ I1_{ZCR},\ I1_{SS},\ I1_{SF},\ I2_{En},\ I2_{PSD},\ I2_{STD},\ I2_{ZCR},\ I2_{SS},\ I2_{SF}). \tag{3.22}$$

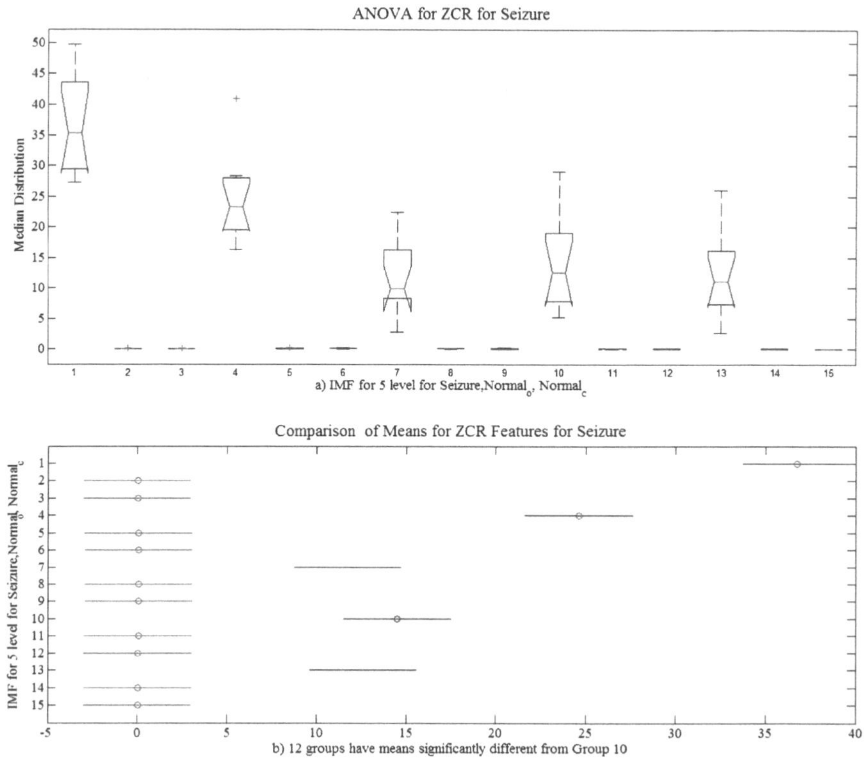

Figure 3.27 Boxplot showing a comparison of ZCR values of IMF1 to IMF5 in the Seizure, Normal_o, and Normal_c groups.

For SSEMD methods, IMF1 and IMF2 show significance after applying ANOVA to the generated decomposed IMFs. Thus all the discussed features were extracted and ANOVA was performed. It shows the effective features for SSEMD features method are PSD, ZCR, and spectral slope. Thus the feature vector constructed for SSEMD is

$$F6 = f(I1_{PSD},\ I1_{ZCR},\ I1_{SS},\ I2_{PSD},\ I2_{ZCR},\ I2_{SS}). \qquad (3.23)$$

Feature vectors F1, F3, F5, and F6 are submitted as input to the classifiers for improved seizure detection.

3.4 Classification methods

The classification algorithms are used to decide to which category the recorded EEG signal belongs, i.e., normal or epileptic. This step includes a strategy for determining prominent features to be selected and constructing a combined feature vector to optimize performance.

The machine learning classifiers can improve the performance and are evaluated by cross-validation. This statistical method works on the basis of data sampling to estimate the classification model. It randomly divides available feature data into training and testing subsets. Methods used in this work are k-fold cross-validation and 66% split cross-validation for the naive Bayes classifier, the multiclass classifier, and the decision tree-based J48 classifier for performance evaluation.

The naive Bayes classifier is a Bayesian theory-based probability classification method used to handle multiclass classification problems. The advantage of this classifier is that a small set of the attribute is sufficient to estimate the class of data. Most of the data mining, machine learning applications adopt this classifier for classification problems. The classifier model is stated as

$$P(X|Y) = P(Y|X) \times P(X)/P(Y), \quad (3.24)$$

where P(X) is the prior probability or marginal probability of X and P(X | Y) is the conditional probability of X. Given Y, called the posterior probability, P(Y | X) is the conditional probability of Y given X. P(Y) is a prior or marginal probability of Y which acts as a normalizing constant. This probabilistic classifier is highly scalable and linear. The multiclass classification problem is used to classify the instances into more than three classes. In machine learning this classifier is used with "N" classes and it can also predict the class label during testing.

The C4.5 classifier works using a decision tree-based methodology and follows the ID3 algorithm. It is a statistical method of classification based on information entropy. J48 is the JAVA implementation of the C4.5 algorithm. Here the highest information gain attribute is selected for the decision of classes, and this is the best-established classifier for the detection and prediction of diseases such as diabetes [24].

The TFD, wavelet, and EMD-IMF-based features are classified using naive Bayes, multiclass, and J48 classifiers with 10-fold validation and 66% split modes of classification.

The classifiers used in the experimental work are given below.

1. Naive Bayes classifier: A naive Bayes classifier is a probabilistic algorithm that uses Bayes' theorem to classify objects. This classifier considers the strong, or naive, independence between attributes of data

points. The naive Bayes classifiers are used for spam filters, text analysis, signal segmentation, and medical diagnosis. It is a preferable method in machine learning because it is simple to implement. The predictive model is defined as

$$p(C_k|x_1 \ldots . x_n) = p(C_k) \prod_{i=1}^{n} p(x_i|C_k), \tag{3.25}$$

where C_k is the class and x_i is the sample of a dataset classified using conditional probability.

2. Multiclass classifier: In machine learning, multiclass or multinomial classification is the problem of classifying instances into one of three or more classes. Classifying events in two classes is called binary classification. Some of the classification methods are used for more than two classes; thus, they are multimodal classifiers used with a variety of strategies. Multiclass classification is different from multilabel classification, which uses multiple labels at each predicted instance.
   Suppose we knew the density, pi(x), for each of the N classes. Then, we would predict using

$$f(x) = arg\ max\ pi(x),$$

   where $i = 1, \ldots, N$. The densities are not known, but can be estimated using classical techniques.
3. J48 Classifier: C4.5 (J48) is an algorithm used to generate a decision tree developed by Ross Quinlan mentioned earlier. C4.5 is an extension of Quinlan's ID3 algorithm. The J48 classifier is a statistical classifier based on decision trees. The WEKA tool implements the J48 algorithm for classification and prediction in precise form.

3.5 Hardware architecture for seizure detection

Now the EEG signal database is used for further feature extraction, feature selection, dimensionality reduction of the feature vector, and classification into the seizure and normal groups. Fig. 3.28 shows the proposed flow for dimensionality reduction of feature vectors. Initial feature extraction and selection is implemented using MATLAB®; the features are input into the EEG processor architecture implemented on field-programmable gate array (FPGA).

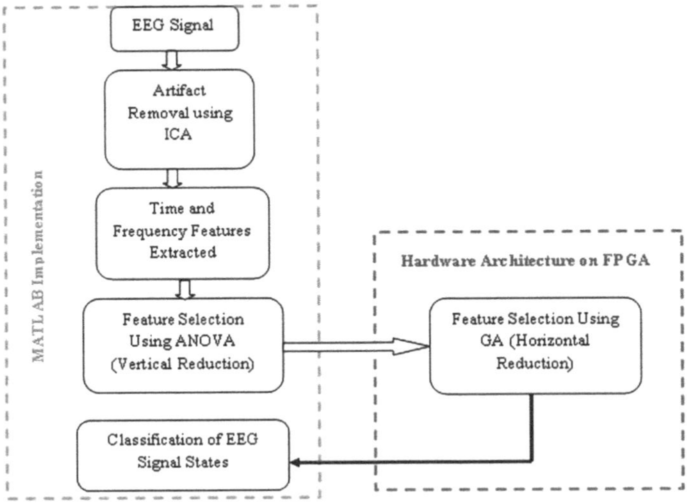

Figure 3.28 Design flow of EEG processing using GA.

3.5.1 Genetic algorithm for dimensionality reduction

Prof. John Holland in 1995 proposed the genetic algorithm (GA), which is supported by Darwin's "survival of the fittest" principle. GA is a natural evolutionary theory used for search and optimization techniques. Such an approach gives very effective solutions for complex problems with high computational demands. The basic GA is implemented in FPGA with crossover and parallel processing in various research. It increases the speed of execution of the algorithm as compared to software implementation. Some features such as accuracy due to hardware, speed of operation, improved search space, and the performance of the floating-point fitness function with general-purpose architecture facilitate FPGA implementation of GA. Thus it is useful in real-time applications, as it is characterized by low power requirements, low size, and high speed [25].

1. **Concept of the genetic algorithm**

   GAs are the search and optimization techniques which follow the process of natural evolution. Such an approach gives very efficient solutions for a complex problem. It keeps improving performance to reach some optimal point [26]. Fig. 3.29 shows the flow of GA. Firstly, an initial population is selected and then the best fitted optimum solution is evaluated for a new population.

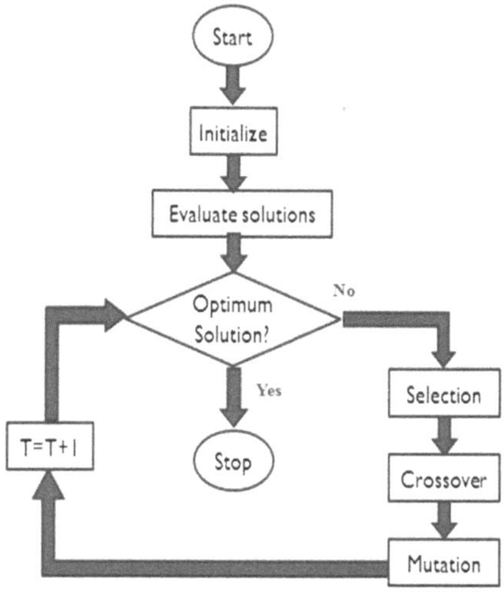

Figure 3.29 Flowchart of a simple GA.

$$\begin{bmatrix} v_{1,1} \cdots v_{1,j} \cdots v_{1,n} \\ \vdots \\ \boxed{v_{i,1} \cdots v_{i,j} \cdots v_{i,n}} \\ \vdots \\ v_{l,1} \cdots v_{l,j} \cdots v_{l,n} \end{bmatrix} \leftarrow i\text{-th gene}$$

Figure 3.30 Genetic coding of a test pattern.

2. **The genetic coding**
   A single EEG result pattern is considered as individuals (or chromosomes) in the GA. Their genetic coding, described below, is a matrix of logic values. Let V_i be an input vector at time point *i*, $V_i = [v_1, \ldots, v_n]$, where n is the number of input signals, and the "v" values are the respective values. An EEG result pattern is a sequence $[V_1, \ldots, V_l]$ of consecutive input vectors, where l is the length of the pattern. Therefore, a pattern is represented by a matrix of size $l \times n$, as shown in Fig. 3.30. The *i*-th index of row represents the gene corresponding to the input vector Vi applied at time point *i*. The *j*-th column corresponds to the sequence of values on the *j*-th input. It may be noted that the number of genes in chromosomes is not assumed to be constant since the time point can take a different value for each EEG result pattern.

3. **Selection**

   A selection operator chooses a subset of chromosomes from the current population. Various stochastic selection techniques are available. In this work, the roulette wheel method is used. With this method, an individual is selected by using the fitness function with probability. Each individual is mapped to an arc of a circle whose length equals the individual fitness. The circumference is then equal to the sum of the fitnesses. Selection is performed by choosing a random number with a uniform distribution between 0 and the circumference. The selected individuals are those individuals which satisfy fitness criteria. It ensures that better fit individuals have a greater probability of being selected; however, all individuals have a chance.

4. **Fitness function**

   The fitness function improves the quality of the solution, and for all populations, it will check the probability of their fitness. In our approach, fitness takes into account the fault coverage achieved by each EEG pattern, its length, and its effectiveness in finding hard seizures. The fitness function adopted in this work relies on a dynamic global record table to evaluate each EEG pattern with respect to the performance of the previously generated EEG pattern. The fitness function of an EEG pattern is defined in terms of a value that we call the relative efficiency of the EEG pattern.

5. **Crossover function**

   The basic operator for producing new chromosomes in the GA is that of crossover. Like its counterpart in nature, crossover produces new individuals that have some parts of both parents' genetic material. The simplest form of crossover is that of single-point crossover, described in the overview of GAs. In this section, a number of variations of crossover are described and discussed, and the relative merits of each are reviewed.

3.5.2 Genetic algorithm implementation on FPGA

1. **FPGA implementation of GA**

   The processor is designed with 10-byte transmission and reception in one clock cycle, and the byte is first stored in block RAM. One frame is of 512 bytes, and then the processing starts in an EEG–GA processor (EEG-GAP) architecture [25]. These frames are categorized as selected features or discarded from memory.

2. **FPGA implementation of UART**

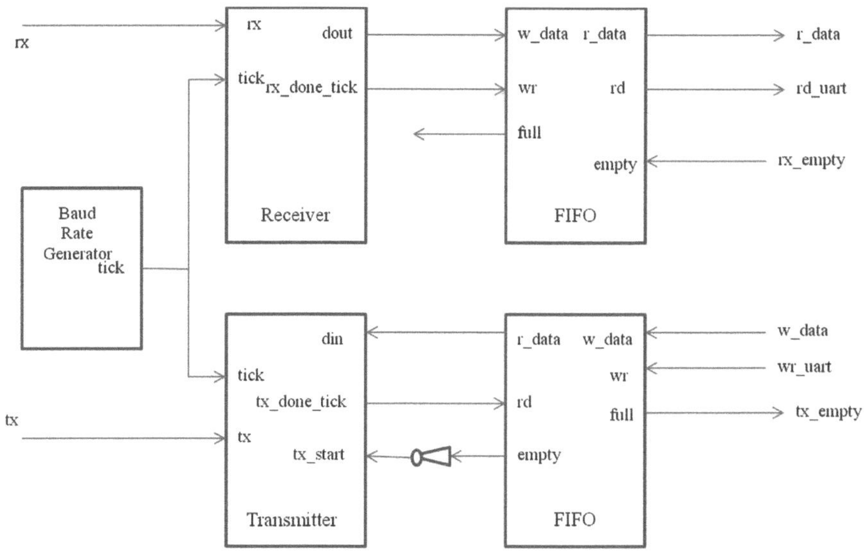

Figure 3.31 UART model implemented on FPGA.

The UART model design is shown in Fig. 3.31. Features from the MATLAB tools are transmitted through RS 232 and the hyperterminal of PC to FPGA. The designed is verified by implementation on a Xilinx Spartan 6 FPGA-based Altys board.

Fig. 3.32 shows a simulation of UART, demonstrating transfer of data. It is divided in three main parts, viz. a communication module for transferring features to FPGA, receiving reduced features from FPGA to PC through RS 232 and UART communication using 8-bit frame size with 8-bit buffers. Harpale et al. [27] published a UART architecture with buffer implementation on FPGA and the work was further extended for dimensionality reduction using GA.

3. **Overall FPGA architecture for EEG-GAP**

The datapath architecture includes clock generation, reading buffer into block RAM, processing through the GA process with a separate process control unit, and uploading the processed data with transmitting buffer to PC. How good a solution is (chromosome) is determined by the fitness criteria for the chosen problem. Generally, the better a solution, the higher is its fitness value. Evaluation is performed by the evaluation module, which computes fitness for each member of the generated population. The computed fitness values for each member

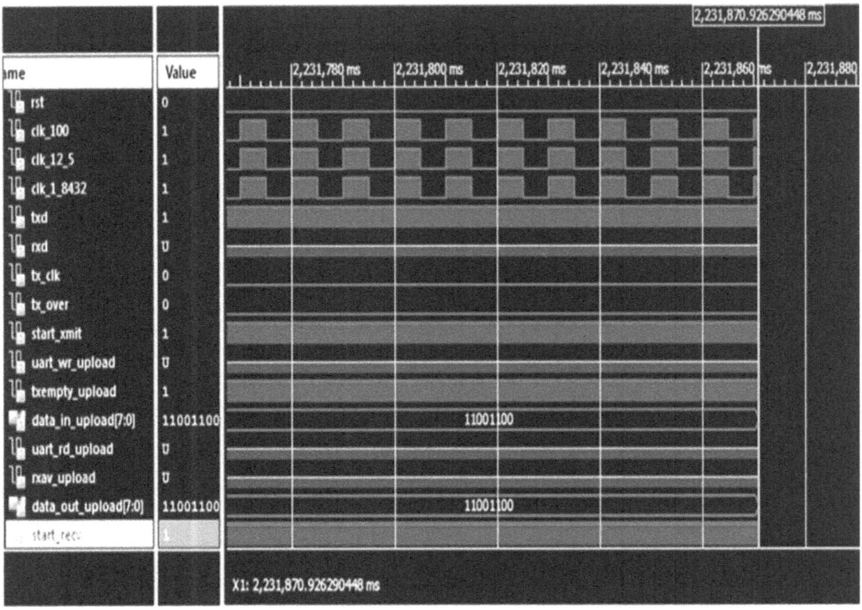

Figure 3.32 Simulation of UART.

are stored in the fitness memory corresponding to the address of its chromosome in the main memory.

The overall architecture is shown in Fig. 3.33. The architecture is designed with DATAPATH and CONTROL blocks. EEG-GAP essentially consists of a processing unit which has modules for performing basic genetic operations. The input to GAP is an encoded chromosome. The coding mechanism can be real numbers or simple binary or backed BCD, depending on the problem. A chromosome is any random solution to the problem. This chromosome is then used for generating a random population of a predefined size. The population of chromosomes represents a set of possible solutions for a given problem. The initial population is stored in the main memory.

It includes input buffer, block RAM, an EEG-GA processor block, and an output buffer. It reads features from the PC sequentially, and serially data will be received in the input buffer and block RAM will be enabled until it receives 512 bytes as one frame. Then the EEG-GA process is initiated to calculate its significance. The total block operations are controller by a control block, which has an 8-bit control word

DB7	DB6	DB5	DB4	DB3	DB2	DB1	DB0
Finish	0	S/E Xmit	S/E up	S/E Proc	S/E Dwn	S/E Rec	SW0

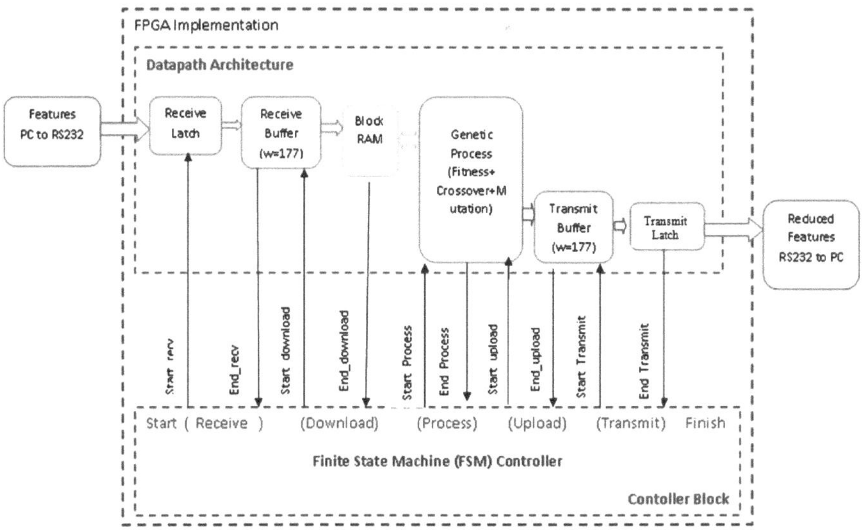

Figure 3.33 Overall architecture of EEG-GAP.

with the following details: **Control Word** The optimum feature is a result of GA by using a fitness function. Here the fitness function is used to find the fitness of the Seizure feature, which has a high PSD. Fig. 3.34 shows architectural details of each of the block. It shows signal details, control, and the datapath distribution for the GA architecture. The blocks are invoked or controlled by a controller designed using a finite state machine (FSM), as shown in Fig. 3.35. The FSM controller uses a control word given below to control the total process execution. The datapath is designed to handle the flow of data. It includes input buffer, block RAM, an EEG-GA processor block, and an output buffer. It reads features from PC sequentially, and serially data will be received in the input buffer. Block RAM will be enabled until it receives 174 bytes as one frame and then the EEG-GA process is initiated to calculate its significance. The blockwise FPGA implementation results are observed using register-transfer level (RTL) schematics, as shown in Figs. 3.36, 3.37 and 3.38.

The overall simulation is shown in Fig. 3.39. The implementation setup is shown in Fig. 3.40. The device utilization summary for the Xilinx-

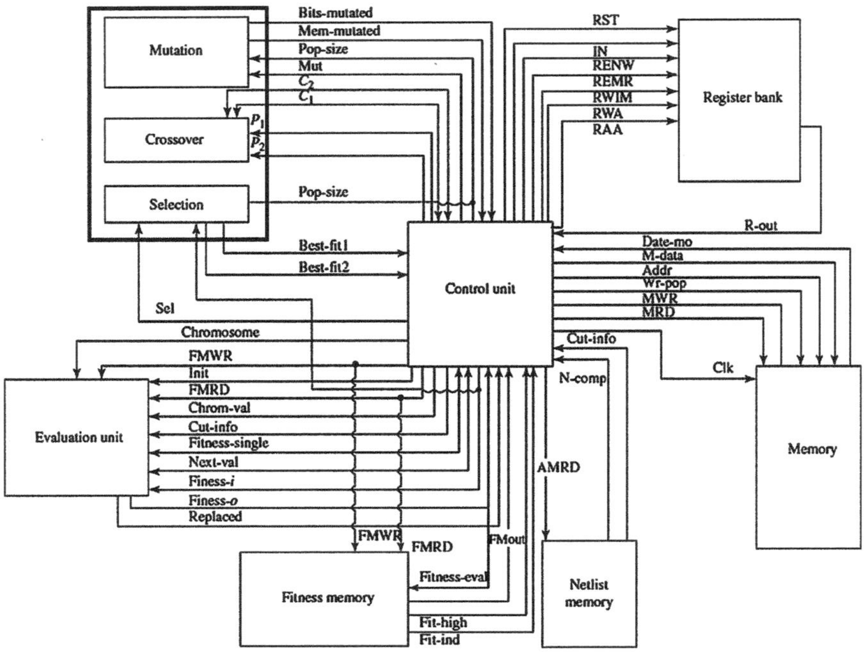

Figure 3.34 Architectural details of EEG-GAP.

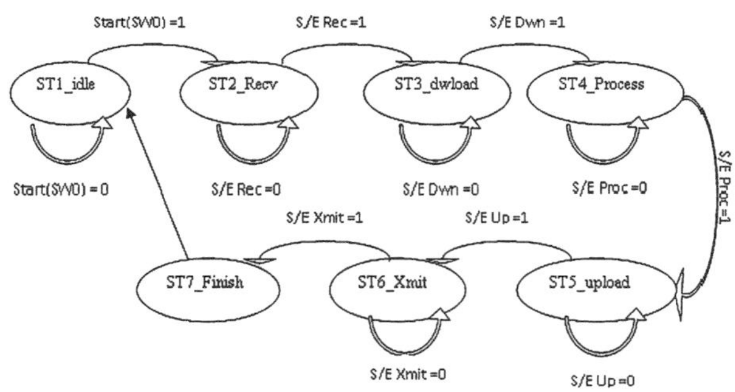

Figure 3.35 FSM controller for EEG-GAP.

SPARTAN 6 FPGA platform is shown in Table 3.2 with 37 mW and an execution time of 18.76 seconds. Features selected from MATLAB are passed through this design and the dimension of the overall features related to the EEG signal in the seizure state is reduced. The significant features related to the seizure state are selected and transmitted to the PC. Thus the reduced set of features is employed to detect seizures. The

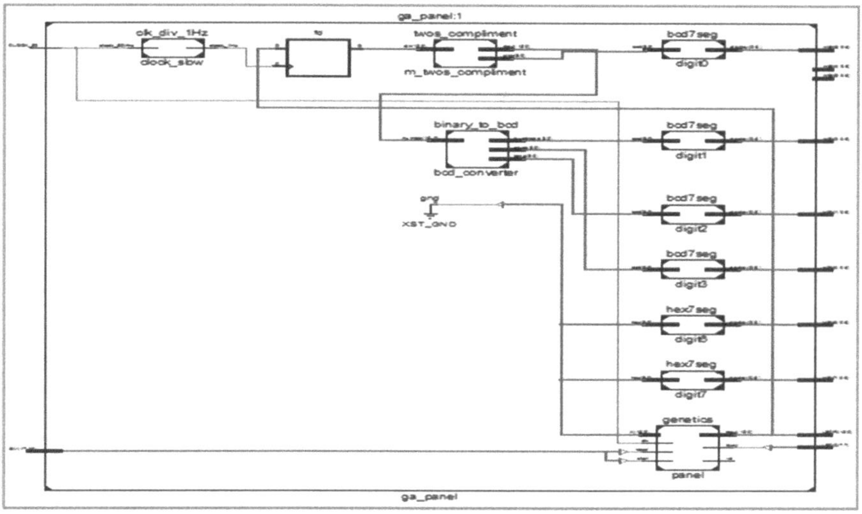

Figure 3.36 RTL schematic of overall architecture of EEG-GAP.

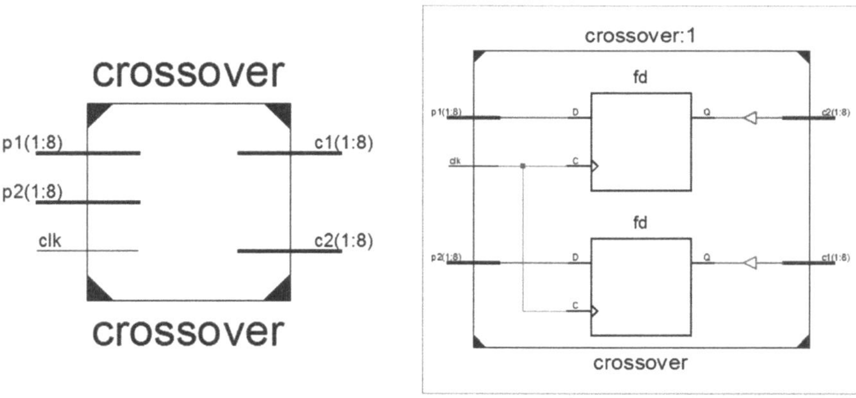

Figure 3.37 RTL simulation of crossover architectural block.

seizure detection with this methodology exhibits improved accuracy and sensitivity. The hardware architecture facilitates its use in real-time ASIC solutions, but performance related to the speed of execution and computation cost is platform-specific. The optimized features are classified using a support vector machine classifier. The WEKA tool is a GUI-based machine learning tool used to evaluate the performance of

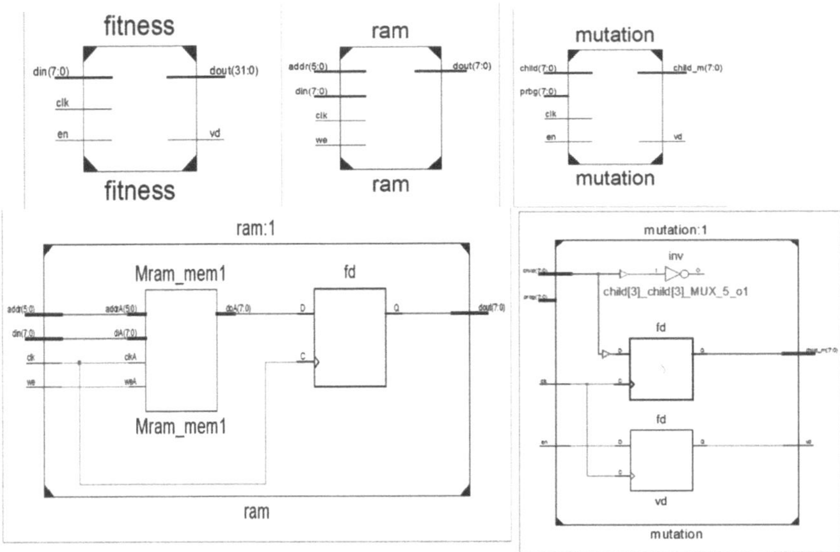

Figure 3.38 Blockwise RTL schematic of fitness and mutation function with RAM module.

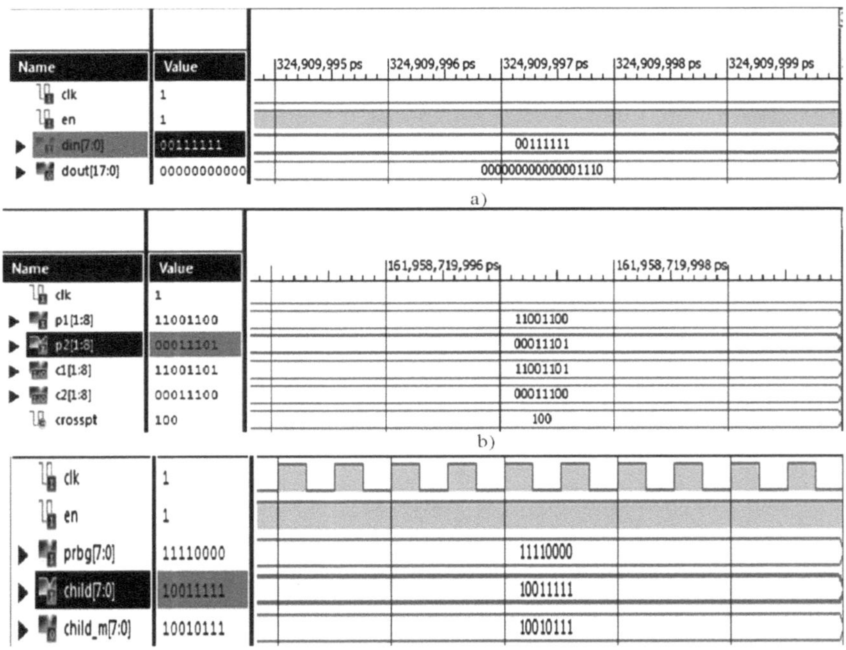

Figure 3.39 Overall simulation of the EEG-GA processor.

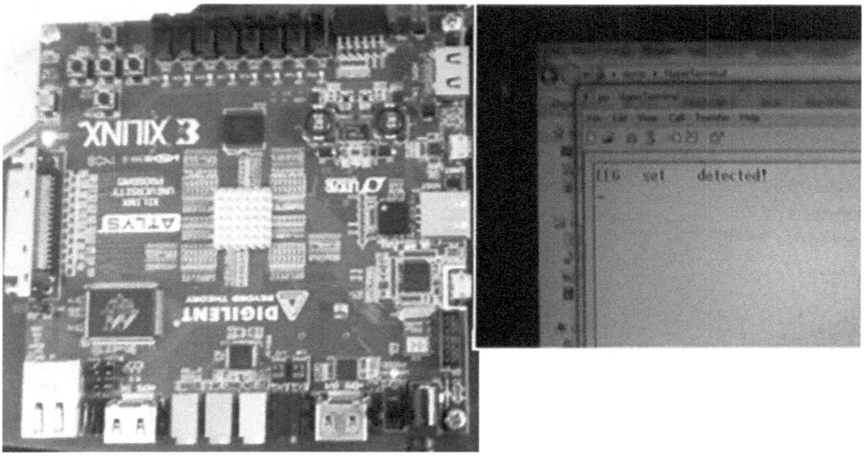

Figure 3.40 Implementation of EEG-GAP using a SPARTAN 6 board.

Table 3.2 Device utilization summary of FPGA implementation.

Logic utilization	Devices used	Devices	Device utilization (%)
Number of slice registers	620	54576	1
Number of slice LUTs	1524	27288	5
Number of fully used LUT-FF pairs	555	1589	34
Number of bonded IOBs	62	218	28
Number of block RAM/FIFOs	13	116	11
Number of BUFG/BUFGCTRLs	1	15	6
Number of DSP 48As	13	58	22

the classifier. The performance metric observed is mentioned in Table 3.3.

3.6 Result and performance analysis of seizure detection

As discussed earlier, the methodology adopted for research is feature extraction and features selection using ANOVA. With the application of optimized classifiers, the system is implemented to achieve good accuracy.

3.6.1 Seizure detection results

TFD, wavelet, and EMD-IMF-based features are classified using various classifiers such as naive Bayes, multiclass, and J48 with 10-fold and 66% split

Table 3.3 Performance parameters of the presented method.

Performance methods	Feature space	% reduction in feature space	Classification accuracy
Time and frequency domain features	23×100×3×5 = 6900×5 features	NA	87.44%
Feature selection using ANOVA	23×100×3×3 = 6900×3 features	40% reduction	92.81%
Optimized ANOVA+FPGA-based GA	[460 (nonseizure features) + 2070 (seizure features)] × 3 = 2530 × 3 (considered maximum value)	63% reduction	95.94%

modes of classification. The original dataset is partitioned into a training set and a test set to evaluate the predictive model; this method is called cross-validation.

In k-fold cross-validation, the dataset is randomly partitioned into k subsets of equal size. The classifiers used in this work employ 10-fold cross-validation for randomly partitioning the feature set into a test set and a training set to verify its performance.

The split validation is a way to predict the fit of a model to a hypothetical test set when an explicit test set is not available. The split validation operator also allows training on one dataset and testing on another explicit test dataset. The present work uses a split ratio of 67% for training and 33% for testing.

The performance parameters are the following.

1. Sensitivity: Sensitivity is defined as the proportion of seizures that are accurately detected:

$$Sensitivity(Sv) = \frac{S_d}{S_t}, \tag{3.26}$$

where S_d is the number of accurately predicted seizures and S_t is the total number of seizures observed during the same interval of time.

2. False positive rate: The false positive rate is defined as the number of seizures per hour that the algorithm does not detect:

$$FPR = \frac{S_t - S_d}{T}. \tag{3.27}$$

3. Classification accuracy: Accuracy is defined as the ratio of correctly classified seizures of all classified seizures for that time interval:

Table 3.4 Performance Parameters Definitions.

Sr. No.	Parameters
1.	True Positive Rate (TPR)/Recall = TP/(TP+FN)
2.	False Positive Rate (FPR) = FP/(FP+TN)
3.	Precision (P)= TP / (TP+FP)
4.	F-Measure(F1) =(2*(P*TPR))/((P+TPR))
5	ROC = Recall Vs FPR Curve

$$Accuracy = \frac{TP}{TP + FP}, \tag{3.28}$$

where TP is the number of true positives and FP is the number of false positives.

4. Specificity: Specificity is the ratio of the number of true negative decisions to the total number of negative cases.

Thus this work aims to achieve the highest classification accuracy, maximum sensitivity, and minimum FPR. This work used the Normal_o, Normal_c, and Seizure EEG signals for experimentation.

The performance parameters observed by the execution of the proposed methodology using feature extraction, feature selection, and machine learning-based classifiers are stated in Table 3.4. Various iterations were performed to optimize performance. The performance metric was constructed, and the following classification performance parameters were measured: true positive rate, false positive rate, precision, recall, F-measure, and receiver operating characteristic (ROC), which is calculated as specified in Table 3.4 on the basis of the following test statistics parameters:

- true positive (TP): the number of events correctly detected,
- true negative (TN): the number of normal events correctly detected,
- false positive (FP): the number of nonexisting events incorrectly identified,
- false negative (FN): the number of existing events detected as normal.

The performance was observed based on TFD feature extraction, and it is observed that the TPR is very good and FPR is moderate, whereas other parameters can be improved. The feature extraction performance of wavelet decomposition shows that the TPR is improved, but other parameters need to improve. It also shows that EMD-based performance is superior to TFD and wavelet transform with respect to the false positive rate (FPR). By observing these results, a feature vector is reconstructed using TFD features and EMD features to obtain improve the TPR and FPR.

Table 3.5 Performance analysis for TFD, wavelet transform (WT), EMD, and SSEMD using machine learning classifiers. A, 10-fold. B, 66% split.

Method and classifier	TPR	FPR	*P*	F1	ROC
TFD (A)					
Naive Bayes	0.958	0.009	0.981	0.97	0.998
Multiclass	0.981	0.015	0.97	0.976	0.998
J48	0.984	0.014	0.97	0.978	0.996
TFD (B)					
Naive Bayes	0.976	0.011	0.978	0.977	0.997
Multiclass	0.984	0.016	0.968	0.976	0.998
J48	0.992	0.017	0.966	0.979	0.992
WT (A)					
Naive Bayes	0.916	0.014	0.97	0.942	0.993
Multiclass	0.95	0.011	0.978	0.964	0.995
J48	0.944	0.016	0.967	0.955	0.981
WT (B)					
Naive Bayes	0.899	0.011	0.977	0.936	0.992
Multiclass	0.94	0.009	0.981	0.96	0.993
J48	0.945	0.015	0.971	0.958	0.978
EMD (A)					
Naive Bayes	0.864	0.004	0.99	0.923	0.976
Multiclass	0.955	0.007	0.986	0.97	0.995
J48	0.956	0.002	0.96	0.958	0.976
EMD (B)					
Naive Bayes	0.863	0.001	0.997	0.925	0.979
Multiclass	0.964	0.003	0.994	0.979	0.997
J48	0.969	0.002	0.943	0.956	0.967
SSEMD (A)					
Naive Bayes	0.96	0.008	0.984	0.972	0.996
Multiclass	0.991	0.001	0.997	0.994	1
J48	0.988	0.007	0.987	0.987	0.992
SSEMD(B)					
Naive Bayes	0.972	0.007	0.986	0.979	0.997
Multiclass	0.998	0.001	1	0.994	0.999
J48	0.992	0.008	0.983	0.987	0.993

Table 3.5 presents the final experimental results. The seizure detection accuracy is observed to be 99.8%, with a false positive rate of zero with 66% split cross-validation and the multiclass classifier.

Table 3.6 Dimensionality reduction using one-way ANOVA.

Method	Feat. Selected Dim.	% Reduction
TFD	3×11 windowed samples ×100 files = 3300 instances	72.72%
WT	8×11 windowed samples ×2 freq. bands×100 files = 17,600 instances	70.9%
EMD, SSEMD	8×11windowed sample ×2 frequency bands ×100 files = 17,600 instances	70.9%

Table 3.7 J48: TFD 10-fold cross-validation.

Class	TP rate	FP rate	Precision	Recall	F-measure	ROC area
Seizure	0.984	0.014	0.972	0.984	0.978	0.996
Normal_o	0.89	0.059	0.883	0.89	0.886	0.952
Normal_c	0.854	0.063	0.871	0.854	0.862	0.936

Table 3.8 J48: TFD split 66.0% training.

Class	TP rate	FP rate	Precision	Recall	F-measure	ROC area
Seizure	0.992	0.017	0.966	0.992	0.979	0.992
Normal_o	0.869	0.048	0.906	0.869	0.887	0.951
Normal_c	0.868	0.071	0.854	0.868	0.861	0.933

One-way ANOVA used for feature reduction and thus dimensionality reduction helps in improving the performance of the model. Table 3.6 shows the overall reductions in features and complexity, improving the performance.

The EEG signal is classified as a Seizure signal, a Normal_o EEG signal, or a Normal_c EEG signal. The experimented results for TFD features are shown in Tables 3.7–3.11. It shows performance parameters observed using TFD feature extraction and J48, multiclass, and naive Bayes classifiers. The multiclass classifier with 66% cross-validation shows good performance, i.e., 98.4%, 85.9%, and 84.3% for Seizure, Normal_o, and Normal_c, respectively.

For epileptic seizure detection, invasive and noninvasive EEG signals are analyzed, achieving 98% and 96.48%, respectively. TPR improved with wavelet transform as compared to TFD, but the FPR improved with the EMD method. The hybrid model of TFD, wavelet transform, and EMD

Table 3.9 TFD: Multiclass 10-fold cross-validation.

Class	TP rate	FP rate	Precision	Recall	F-measure	ROC area
Seizure	0.981	0.015	0.97	0.981	0.976	0.998
Normal_o	0.849	0.066	0.866	0.849	0.857	0.967
Normal_c	0.839	0.085	0.832	0.839	0.836	0.942

Table 3.10 TFD: Multiclass: split 66.0% training.

Class	TP rate	FP rate	Precision	Recall	F-measure	ROC area
Seizure	0.984	0.016	0.968	0.984	0.976	0.998
Normal_o	0.859	0.061	0.881	0.859	0.87	0.968
Normal_c	0.843	0.08	0.834	0.843	0.839	0.942

Table 3.11 TFD: Naive Bayes classifier: 10-fold cross-validation.

Class	TP rate	FP rate	Precision	Recall	F-measure	ROC area
Seizure	0.958	0.009	0.981	0.958	0.97	0.998
Normal_o	0.855	0.25	0.63	0.855	0.726	0.925
Normal_c	0.841	0.094	0.72	0.481	0.577	0.9

improves seizure detection to 98% and the false positive rate to 0.3% for the classification of controls and epileptic patients without observing seizure onset, and could lead to the early detection of epilepsy.

3.7 Summary

For EEG analysis and seizure detection, feature extraction and feature selection are important requirements to avoid high-dimensional spaces for data management. For epileptic seizure detection, invasive and noninvasive EEG signals are analyzed. Seizure occurrence can be further analyzed for early detection of seizures. The feature dimensionality reduction is about 71% using SSEMD and ANOVA and 63% using the GA-ANOVA approach, thus reducing the computational complexity of the classifier. The system will be significant in seizure monitoring of epileptic patients as well as detection of seizures to avoid accidents. In conclusion, the basic EEG-GA architecture exhibits high performance for seizure detection and can be further extended to seizure prediction.

References

[1] V.K. Harpale, V.K. Bairagi, Time and frequency domain analysis of EEG signals for seizure detection, in: Proceeding of the IEEE International Conference on Microelectronics, Communication & Computing (MicroCom 2016), NIT, Durgapur, 2016.
[2] Andrea Varsavsky, Iven Mareels, Mark Cook, Epileptic Seizures and the EEG Measurement, Models, Detection and Prediction, CRC Press, 2011.
[3] R.G. Andrzejak, K. Schindler, C. Rummel, Nonrandomness, nonlinear dependence, and nonstationarity of electroencephalographic recordings from epilepsy patients, Physical Review. E 86 (2012) 046206.
[4] Thasneem Fathima, M. Bedeeuzzaman, Omar Farooq, Yusuf U. Khan, Wavelet based features for epileptic seizure detection, MES Journal of Technology and Management 2 (1) (2011) 108–112.
[5] R. Jenke, A. Peer, M. Buss, Feature extraction and selection for emotion recognition from EEG, IEEE Transactions On Affective Computing 5 (3) (2014) 327–339.
[6] H.K. Sawant, Z. Jalali, Detection and classification of EEG waves, Oriental Journal of Computer Science and Technology 3 (1) (2010) 207–213.
[7] M. Rosenblatt, A. Figliola, G. Paccosi, E. Serranoand, O.A. Rosso, A quantitative analysis of an EEG epileptic record based on multi resolution wavelet coefficients, Entropy 16 (11) (2014) 5976–6005.
[8] S. Nasehi, H. Pourghassem, An optimal EEG-based emotion recognition algorithm using Gabor, WSEAS Transactions on Signal Processing 3 (8) (2012) 87–99.
[9] M. Maeda, A. Takajyo, K. Inoue, Kousuke Kumamaru, S. Matsuoka, Time-frequency analysis of human sleep EEG and its application to feature extraction about biological rhythm, in: Proceeding of SICE, Annual Conference, 2007, pp. 1939–1944.
[10] O. Faust, U.R. Acharya, H. Adeli, A. Adeli, Wavelet-based EEG processing for computer-aided seizure detection and epilepsy diagnosis, Seizure 26 (2015) 56–64.
[11] D. Labate, F.L. Foresta, G. Occhiuto, F.C. Morabito, A. Lay-Ekuakille, P. Vergallo, Empirical mode decomposition vs. wavelet decomposition for the extraction of respiratory signal from single-channel ECG a comparison, IEEE Sensors Journal 13 (7) (2013) 2666–2674.
[12] Norden Huang, Zheng Shen, Steven Long, M.L.C. Wu, Hsing Shih, Quanan Zheng, Nai-Chyuan Yen, Chi-Chao Tung, Henry Liu, The empirical mode decomposition and the Hilbert spectrum for nonlinear and non-stationary time series analysis, Proceedings of the Royal Society of London. Series A: Mathematical, Physical and Engineering Sciences 454 (1998) 903–995, https://doi.org/10.1098/rspa.1998.0193.
[13] Zeiler Angela, Faltermeier Rupert, Keck Ingo, Tomé Ana, Puntonet Carlos, Lang Elmar, Empirical mode decomposition – an introduction, in: Proceedings of the International Joint Conference on Neural Networks, 2010, pp. 1–8.
[14] Sabrina Belhadj, Arwa Attia, Bachir Ahmed Adnane, Zoubir Ahmed-foitih, A novel epileptic seizure detection using fast potential-based hierarchical agglomerative clustering based on EMD, International Journal of Computer Science and Network Security 16 (5) (2016) 21–29.
[15] R. Sharma, R.B. Pachori, U.R. Acharya, Application of entropy measures on intrinsic mode functions for the automated identification of focal electroencephalogram signals, Entropy 17 (2015) 669–691.
[16] Vinayak K. Bairagi, Varsha K. Harpale, Improved epileptic seizure detection using singular spectrum empirical mode decomposition and machine learning approach, Journal of Statistics and Management Systems (2021), https://doi.org/10.1080/09720510.2020.1862958.
[17] S. Enshaeifar, S. Kouchaki, C.C. Took, S. Sanei, Quaternion singular spectrum analysis of electroencephalogram with application in sleep analysis, IEEE Transactions on Neural Systems and Rehabilitation Engineering 24 (3) (2015) 57–67.

[18] A.K. Maddirala, R.A. Shaik, Motion artifact removal from single channel electroencephalogram signals using singular spectrum analysis, Biomedical Signal Processing and Control 30 (2016) 79–85.
[19] Nai-Fu Chang, Tung-Chien Chen, Cheng-Yi Chiang, Liang-Gee Chen, Channel selection for epilepsy seizure prediction method based on machine learning, in: Proceeding of Annual International Conference of the IEEE Engineering in Medicine and Biology, 2012.
[20] Nadir Omer Fadl Elssied, Othman Ibrahim, Ahmed Hamza Osman, A novel feature selection based on one-way ANOVA f-test for e-mail spam classification, Research Journal of Applied Sciences, Engineering, and Technology 7 (2014) 625–638.
[21] M.O. Arowolo, S.O. Abdulsalam, Y.K. Saheed, M.D. Salawu, A feature selection based on one-way-ANOVA for microarray data classification, Al-Hikmah Journal of Pure and Applied Sciences 3 (2016) 35–42.
[22] Varsha Harpale, Vinayak Bairagi, An adaptive method for feature selection and extraction for classification of epileptic EEG signal insignificant states, Journal of King Saud University – Computer and Information Sciences (ISSN 1319-1578) (2018), https://doi.org/10.1016/j.jksuci.2018.04.014.
[23] V.K. Harpale, V.K. Bairagi, Significance of independent component analysis (ICA) for epileptic seizure detection using EEG signals, in: Proceedings of the International Conference on Data Engineering and Communication Technology (ICCUBEA2017), Pune, in Advances in Intelligent Systems and Computing 2, 2017, pp. 829–838.
[24] Gaganjot Kaur, Amit Chhabra, Improved J48 classification algorithm for the prediction of diabetes, International Journal of Computer Applications 98 (2014) 13–17.
[25] Varsha Harpale, Vinayak Bairagi, FPGA based architecture implementation for epileptic seizure detection using one way ANOVA and genetic algorithm, Biomedical and Pharmacology Journal 12 (3) (2019) 1543–1553.
[26] P.D. Thorabole, V.K. Harpale, V.K. Bairagi, Hardware implementation of genetic algorithm for epileptic seizure detection and prediction, in: Proceeding of the IEEE 3RD International Conference on Computing, Communication, Control and Automation (ICCUBEA 2017), Pune, Aug. 2017.
[27] V.K. Harpale, V.K. Bairagi, S.S. Pujari, Real time SoC architectures for analysis of EEG, in: Proceeding of 3RD IEEE: International Conference on Computing, Communication, Control and Automation, 2017, pp. 1–5.
[28] S. Shukla, S. Dandapat, S.R.M. Prasanna, Spectral slope based analysis and classification of stressed speech, International Journal of Speech Technology 14 (3) (2011) 245–258.

CHAPTER 4

Pattern adapted wavelet-based seizure prediction

4.1 Recent trends in epileptic seizure detection using fuzzy classifiers

The published research distinguishes between normal electroencephalographic (EEG) signals and epileptic seizure signals. The primary objective of the present chapter is to identify the preseizure state by tracking the brain state using time and frequency features. The features are classified by a fuzzy system for prediction of seizures. Also, the methodology uses pattern adapted wavelet transform (PAWT) to refine feature selection and classification to detect seizures and classify signals in the preseizure and seizure states. The research provides a technical system for the prediction of seizures well in advance with an appropriate alert to handle the seizure and avoid accidents. In this section, the PAWT method is used for preseizure detection, and seizures are predicted 110 seconds before the actual arrival.

Epileptic seizure detection is a critical problem that is still actively researched. Basic classifiers mostly used are support vector machine (SVM), K-means clustering, fuzzy, and adaptive neural network (ANN) classifiers. Tang et al. [1] proposed, a SVM-based tunable assembly classifier and achieved an accuracy of 98.72%. Adaptive neural fuzzy networks (ANFNs) also play an essential role in seizure detection, with 85.9% accuracy. Bekir Karlık et al. [2] compared various machine learning methods such as ANNs, naive Bayesian, k-nearest neighbor (k-NN), SVM, and k-Means classification methods. The features are extracted from the discrete wavelet transform and an autoregressive (AR) model. It was found that k-NN, ANN, and SVM exhibit good performance in epileptic seizure detection. Nasser Sadati et al. [3] compared the performances of various neural networks for different applications. Quidwai et al. [4] proposed a new approach for preseizure, seizure, and postseizure classification using the Labview platform and also provided a fuzzy-based portable solution to seizure detection.

Bahareh-Atoufi et al. [5] analyzed multichannel prediction of seizure, preseizure, and normal states of EEG signals. Nonlinear dynamics were used to select predictor channels. To achieve good accuracy, sensitivity, and

Brain Seizure Detection and Classification Using EEG Signals
https://doi.org/10.1016/B978-0-32-391120-7.00009-8

specificity, most relevant features need to be considered. Typically, if the number of features is high, the performance is good, but computational complexity increases concomitantly. Thus there is a need to reduce this complexity by selecting the most relevant features. Most of the work is restricted to the classification of seizure and nonseizure EEG signals. This work divides EEG signals into preseizure, interseizure, and postseizure periods to improve performance prediction.

4.2 Dataset

The CHB-MIT EEG scalp dataset from the Children's Hospital in Boston [26] is used in this experimentation. It is an epileptic seizure-based database that consists of scalp EEG recordings from 22 patients with intractable seizures. It is comprised of 5 males and 17 females with ages ranging from 1.5 to 22 years. The recorded EEG data have a sampling rate of 256 samples per second with 16-bit resolution. The international 10–20 system bipolar is used for location of sensors with 23 channels to record the data. The database contains records of 1 to 4 hours, seizure samples are labeled, and at least 30 files are available per subject. Thus, the minimum number of samples used from each single patient is 30 files $\times$ 256 samples/sec $\times$ 60 $\times$ 60 $\times$ 1–4 hours = 27,648,000 to 110,592,000 samples. Some of the files contain more than one seizure instances.

4.3 Feature extraction

4.3.1 Features selected from time and frequency domain

Repeatability, regularity (periodicity), synchronicity, and amplitude variation are measured features which differentiate epileptic seizure signals from normal signals. Time domain and frequency domain characteristics of the signal could define these features. Amplitude variation is analyzed by mean power of the signal, variability and regularity will be calculated by variance, the coefficient of variation (COV), and total variation, periodicity can be justified from autocorrelation with suitable window size, and synchronicity can be estimated from cross-correlation of the signals.

Our approach uses feature vector sets related to amplitude variation and frequency variation. Initially, the system is trained with 12 patients and tested with 10 patients from the database, and then for cross-validation, data from 15 patients are used for training and data from 7 patients are used to test accuracy. The feature vector is constructed using four parameters. The features extracted are discussed in Section 4.1.

As discussed earlier, for the classification of epileptic EEGs, time domain features used include an increase in amplitude, an increase in regularity, and an increase in synchronicity. Average mean, variance, and variability estimates change in EEG signals over time with a suitable window length. Thus time domain features are measure of amplitude variation with time, such as energy, power, mean, and variability, regularity is tested based on variance, and the COV, and the total variation and similarly synchronicity are verified with cross-correlation and phase locking [27].

Various EEG features are used for analysis, but as per the characteristics of EEG seizure signals, their variation between seizure and normal states is tested with analysis of variance (ANOVA), and as discussed feature selection, only some relevant features are selected for classification. Frequency is defined as the occurrence of events in a specified time. Nonstationary EEG signals are comprised of events at different frequencies. The most commonly used frequency domain feature is power spectral density (PSD). The normalized PSD is used to estimate the amount of power caused by events at each frequency. PSD has the limitation that it cannot be used to determine the time at which a frequency of interest occurs.

4.3.2 Feature extraction using pattern adapted continuous wavelet transform

Time and frequency changes during seizure demand a method of feature extraction which addressed these parameters. The wavelet transform has been proven to be an efficient analytical tool for multiscale analysis and is useful in signal processing, pattern recognition and classification, and many more applications. For nonstationary signals like EEG, a mother wavelet function with zero mean and proper localization with frequency and time-space is selected.

In the field of epileptic seizure detection and prediction, discrete wavelet transform serving as a feature extraction method in combination with multiple classifiers such as an adaptive neural fuzzy inference system, a fuzzy neural network (FNN), and fuzzy classifiers was implemented, and good performances have been observed. Some special classifiers such as the ant colony classifier and the statistical pattern recognition-based classifier with DWT were also proven to be good. Continuous wavelet transform (CWT) is a useful method for pattern detection. An approximate wavelet is constructured by extracting the seizure pattern from a single channel of the recording. CWT coefficients are obtained from the PAWT [28].

Features such as average power, the COV, and the root mean square (RMS) error of wavelet coefficients are extracted with the pattern wavelet transform. These features serve as input to the fuzzy classifier to classify signals as preseizure, seizure, and postseizure states [29].

The process used is as follows:

1. Construct a normal EEG signal-based pattern wavelet transform of length 36 centered at 256 samples of the normal signal.
2. Apply PAWT to the signal.
3. Estimate the scale and position of the pattern from absolute values of the wavelet coefficient.
4. The duration of the seizure is the scale sampling period.
5. The centered time is the index sampling period.

Thus seizures are detected with feature vectors constructed using wavelet coefficient and:

1. the mean value,
2. the RMS,
3. PSD, and
4. the standard deviation (σ) of the wavelet coefficient, which is calculated as follows:

$$\sigma = \sqrt{\frac{1}{N}\sum_{i=1}^{N}(x_i - \mu)^2}. \tag{4.1}$$

Fig. 4.1 shows an adapted pattern wavelet scalogram of an EEG signal describing the power distribution, and a frequency component exists in each sample. Each waveform shows a frequency relation and its equivalent energy. PAWT if applied to the EEG signal shows variation in the pattern with its center point and range of variation. The scalogram plotted here is a frequency and time relation of the normal EEG signal for the single channel FP1. A similar procedure was followed for the seizure signal and preseizure signal observed for time and frequency analysis.

4.3.3 Feature selection

The data analysis conducted with EEG signals is a long-time analysis, so generated features for each window length are high in number. Based on maximum relevance and minimum redundant relations, features are selected for further feature vector calculation. Selection of appropriate features is

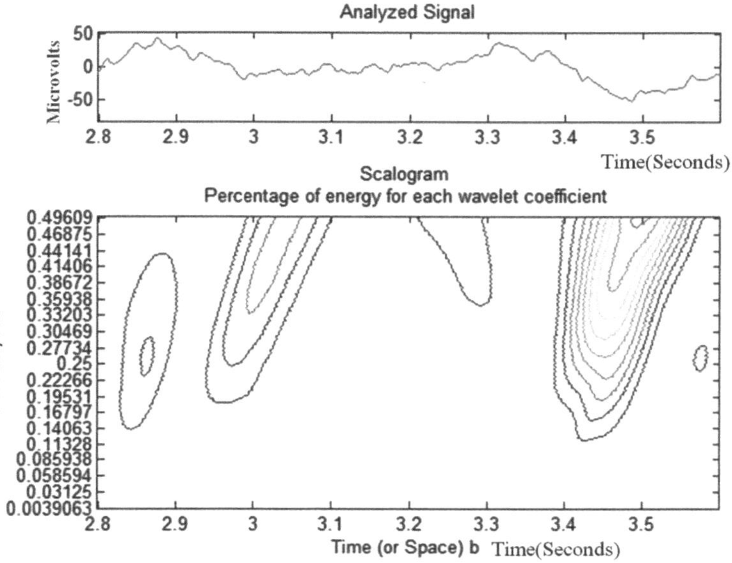

Figure 4.1 Scalogram: Adapted Pattern Wavelet.

essential to reduce computational cost and response time and improve the performance of the system [30]. The fuzzy classifier uses this feature vector for classification of normal, preseizure, and seizure states.

The process of feature extraction and selection used here is described in Fig. 4.2. It shows the first feature vector is constructed using time-frequency features for seizure, normal, and preseizure states. Similarly, a pattern adapted wavelet generates a wavelet coefficient, and features are extracted and selected using ANOVA. The second feature vector is constructed using these features.

4.4 Classification model for preseizure, seizure, and normal states

The random and nonstationary behavior of EEG signals makes the prediction of seizures difficult. Hence seizure detection, prediction, and categorization are very important issues. Various signal processing methods, along with feature extraction, are adopted to categorize the signal segments based on the specific features. Thus there is a need for adaptive seizure detection methods. In complex classifiers and prediction algorithms, some

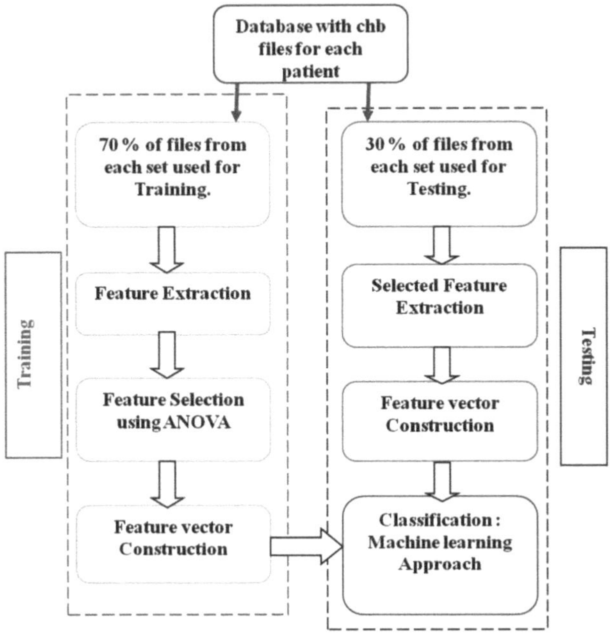

Figure 4.2 Feature extraction and selection flow for classification using fuzzy classifier.

uncertainties can be modeled using random behavior or the fuzzy nature of the input. Various machine learning methods such as ANN, SVM, fuzzy systems, or the recent adaptive neuro-fuzzy inference system classifiers were used to extract multiple features from EEG signals.

4.4.1 Fuzzy logic model

Fuzzy logic is a multivalued logic built upon the theory of fuzzy sets, which was developed by L. Zadeh in 1965. Fuzzy logic resembles human reasoning in its use of approximate information and uncertainty to generate decisions. Systems and models built using fuzzy logic are useful in designing highly complex systems of unknown behavior such as neural systems. Fuzzy logic is used in many applications, ranging from control theory to artificial intelligence. Fuzzy logic includes 0 and 1 as extreme cases of truth; it also consists of the various states of fact, which lie between 0 and 1. Fuzzy logic contains a set of user-supplied human language rules which are further replaced by mathematical equivalents. It maps the real-world behavior with the designed system. Additional benefits of fuzzy logic include its simplicity and its flexibility. Fuzzy logic models, called fuzzy inference systems, consist of several conditional "if-then" rules.

The proposed approach uses two sets of features; one of it is based on simple features calculated from statistical parameters, and others are derived from wavelet transform. The fuzzy model with the Gaussian membership function generates two outputs. Every output describes the state of the signal. Performance was analyzed and found to be improved.

4.5 Feature extraction for preseizure identification

The results presented in this work are calculated from a female patient of the age of 14 with database file chb01_03. The feature vector is constructed from these time domain and frequency domain features. This feature set can be modeled as follows:

$$F(m, n, kavg) = (mean,\ COV,\ RMS,\ PSD), \qquad (4.2)$$
$$\text{with } window = 512\ m <= 22,\ n <= 15,\ k = 23,$$

where "m" is the patient, "n" is a recording of the same patient from the database, and "$kavg$" is the average of features observed from the number of channels used for (m, n).

Feature vectors are calculated from the training dataset. EEG signals of seizures are taken as training reference, and predefined seizure state features are used as trained features during the seizure. It also contains features related to the preseizure state, derived from 30 seconds before the occurrence of the seizure, and other EEG events are used for calculation of features associated with the normal state of the brain. Thus features during the normal state, preseizure state, and seizure state, as shown in Fig. 4.3, are analyzed to find their interrelation. The features are calculated based on the seizure duration specified in the database, and the preseizure state is defined as the 30 seconds before the seizure. The trained time and frequency features are calculated and are shown in Tables 4.1 and 4.2. It shows that the percentage of variation in the features as compared to normal state features can be used as one of the inputs of the fuzzy classifier. These time-frequency features (x1, x2, x3, x4, x5, and x6) are mean, variance, COV, RMS, kurtosis, and PSD, respectively. Maximum frequency calculation will not be considered for classifiers input in this work.

Fig. 4.4(a)–(f) shows a graphical comparison of these features to find prominent features for classification of states in EEG signals. Fig. 4.4(a)

Table 4.1 Summary of recent work in seizure detection using fuzzy classifier variants.

Authors	Methodology	Performance parameters
P. Khosropanah et al. [6]	WT and ANFIS	Sensitivity: 99%
D. Najumnissa et.al [7]	AWT and ANFIS	Accuracy: 97–99%
Sang-Hong Lee et al. [8]	WT and fuzzy, neural network	Sensitivity: 99.67%
A. Subasi et al. [9]	WT and ANFIS	Specificity: 93.7%, sensitivity: 94.3%
Inan Guler et al. [10]	WT and ANFIS	Accuracy: 98.68%
Ahmed F. Rabbi et al. [11]	Fuzzy classifier	High-performance spatial and temporal analysis of EEG signals
D. Gajic et al. [12]	WT, scatter matrices, and quadratic classifier	Accuracy: 99%
Priyanka Jaiswal et al. [13]	EMD and fuzzy	Improved accuracy
Uvais Qidwai et al. [14]	Fuzzy system	Accuracy: 97%
Mohamed Shakir et al. [15]	Intelligent fuzzy classifier	Specificity: 93%, sensitivity: 93.97%, false detection rate: 6.98%
M. Jothi et al. [16]	Fuzzy system on FPGA	Less device utilization
Osman Salem et al. [17]	DWT and ant colony classifier	High detection ratio
B.Pushpa et al. [18]	EMD and ANFIS	Better accuracy
Jing Wang et al. [19]	ANFIS and HS	Accuracy: 97.94%
Meenakshi Sood et al. [20]	Comparative analysis of hardware	Developed a computer-aided diagnosis system
Hamid Niknazar et al. [21]	Fuzzy logic with SBLE	Sensitivity: 100%
C.J. James et al. [22]	Fuzzy rule-based system	Sensitivity: 55.3%, selectivity: 82.0%, false detection rate: 7.2/hr.
Amit K. Sinha et al. [23]	Fuzzy measure-theoretic quantum approximation of an abstract system (FMQAS)	Good accuracy
Gopika Gopan K et al. [24]	Adaptive neuro-fuzzy classifier	Automatic system with 100% accuracy
R. Sukanesh et al. [25]	Structured decision tree in fuzzy optimization	Good performance index

Table 4.2 Features extracted using pattern adapted wavelet transform during normal (N), preseizure (PS), and seizure (S) states.

Feature vector	RMS N	RMS PS	RMS S	Mean N	Mean PS	Mean S	VAR N	VAR PS	VAR S
F(1,1,Kavg)	24.53	54.61	85.16	0.23	0.34	0.15	689	3259	7973
F(1,2,Kavg)	22.88	72.58	91.64	0.18	0.36	0.06	554	6085	9352
F(1,3,Kavg)	36.68	76.03	95.16	0.22	0.32	0.36	1934	6687	10162
F(1,4,Kavg)	23.39	52.9	73.61	0.21	−0.04	0.34	831	3246	6222
F(1,5,Kavg)	33.79	95.22	123.44	0.2	−0.35	0.2	1187	10259	17740
F(1,6,Kavg)	39.34	99.02	131.79	0.2	0.11	0.25	1602	11921	21434
F(1,7,Kavg)	44.41	90.19	139.67	0.21	0.03	0.63	2130	9829	22973

Table 4.3 Features extracted using pattern adapted wavelet transform during normal (N), preseizure (PS), and seizure (S) states continued.

Feature vector	PSD N	PSD PS	PSD S	Kurt N	Kurt PS	Kurt S	COV N	COV PS	COV S
F(1,1,Kavg)	1.68	0.78	1.85	54.45	16.82	6.49	104.3	521.1	0.7
F(1,2,Kavg)	0.95	1.28	2.25	11.8	10.85	6.19	133.9	200.3	−856.1
F(1,3,Kavg)	3.74	2.47	4.17	11.78	7.37	4.91	175.1	312.7	92.5
F(1,4,Kavg)	2.74	1.01	1.8	13.68	12.39	6.12	111.7	−272.6	185.6
F(1,5,Kavg)	1.86	1.31	2.19	58.66	15.66	8.4	191.6	39.1	306.8
F(1,6,Kavg)	2.30	1.25	2.71	34.81	15.05	7.6	201.2	−495.3	−78.7
F(1,7,Kavg)	3.89	1.69	3.88	66.97	15.86	7.1	217	−144.6	104.8

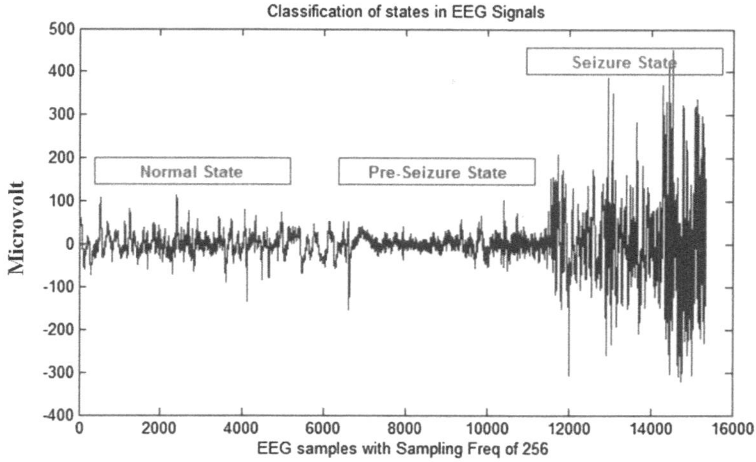

Figure 4.3 Normal, preseizure, and seizure states of EEG signals.

shows time domain features and the mean of the signal, and shows that the mean value of the normal state remains nearly constant as compared to preseizure and seizure states. The mean value in preseizure and seizure states is higher than the normal value, and the highest values are observed in the seizure state. Fig. 4.4(b) shows variance in three states. The behavior of variance is similar to that of the mean, but more distinct and separable.

Fig. 4.4(c) shows the RMS values of the EEG signal. Sufficient variation is observed during the preseizure and seizure states. Fig. 4.4(d) shows that the features related to kurtosis are in line with mean and variance. Fig. 4.4(e) shows the COV observed with the window of 512 samples, but here, no separable relation is observed to consider this feature. PSD is used to estimate features that depend on the frequency and help understanding static properties such as locally stationary and dynamic properties capturing the time-evolving nature. Fig. 4.4(f) shows a graphical plot of the PSD and indicates that it can be used as one of the features to detect seizures, as it has distinct characteristics, and it also shows maximum frequency during the seizure. Results are calculated from EEG signals from chb03 of the CHB MIT database. It is concluded that the mean, variance, RMS, kurtosis, and PSD can be used by the classifier to distinguish between the states in EEG signals.

PAWT if applied to EEG signals shows variation in the pattern with its center point and range of variation. Fig. 4.5 shows adapted pattern wavelet

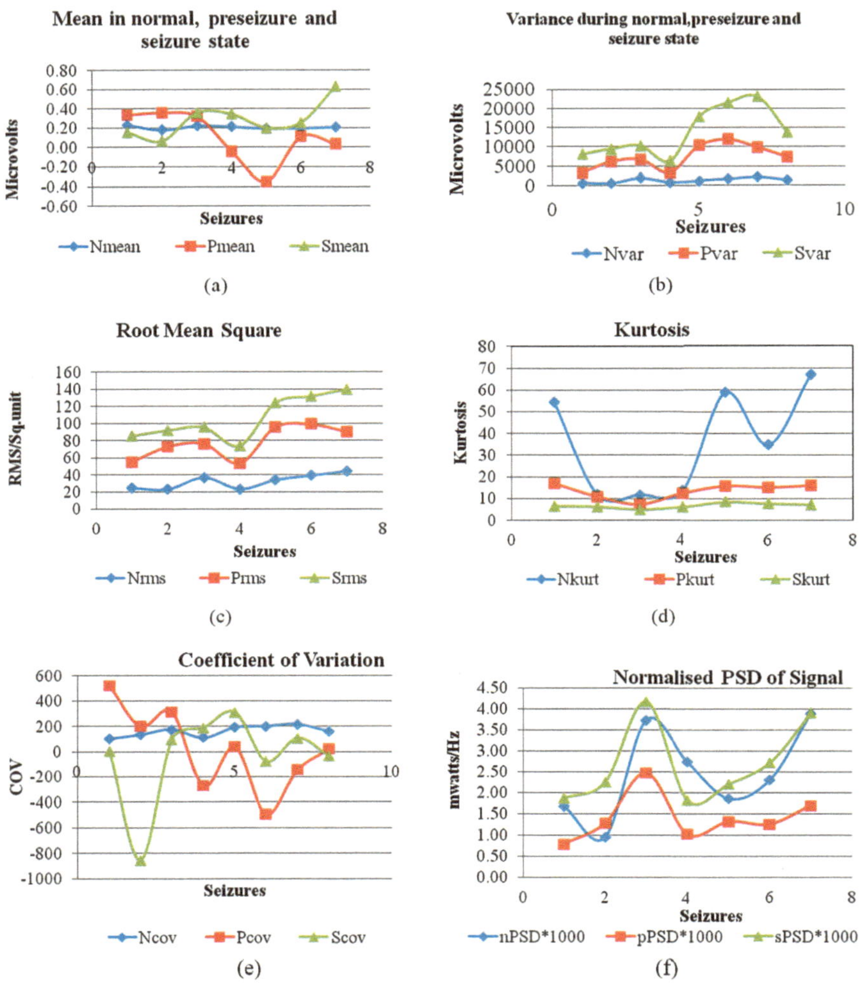

Figure 4.4 Features during normal, preseizure, and seizure states for seven EEG signals containing a single seizure each. (a) Mean. (b) Variance. (c) Root mean square (RMS). (d) Kurtosis. (e) Coefficient of variation (COV). (f) Power spectral density (PSD).

scalograms in the seizure state and preseizure states, and reveals maximum variation of wavelet coefficients. As mentioned, the mean, RMS, PSD, and standard variation of wavelet coefficients are calculated, and feature vectors for these features are calculated.

These features are tabulated in Table 4.2. It is observed that there are variations in features during the preseizure period compared with the nor-

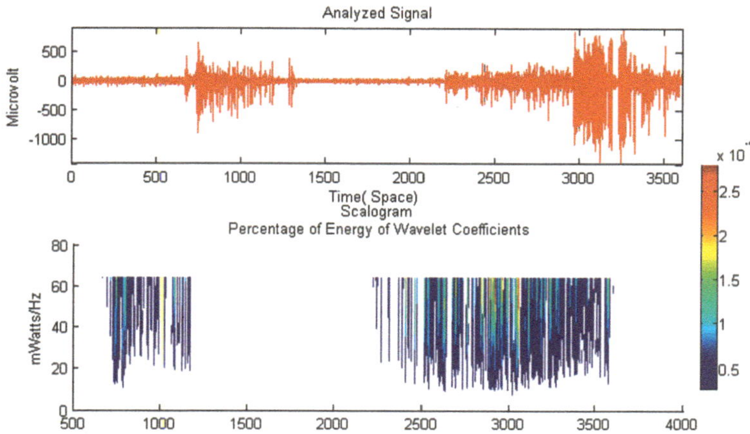

Figure 4.5 Scalogram: adapted pattern wavelet. (a) During seizure. (b) During preseizure.

mal period. Also, there are sudden incremental changes after the preseizure state that are typical of the seizure state. The wavelet-based analysis gives better variation with respect to the features calculated, thus refining the feature extraction phase.

The extracted features are plotted in Fig. 4.6(a)–(d). Fig. 4.6(a) shows the mean value of the coefficient, which has no distinct, separable characteristics. Fig. 4.6(b) shows RMS values during the normal, preseizure, and seizure states and shows separable relations. Fig. 4.6(c) and d shows PSD and standard deviation, respectively, which also show separable characteristics. Therefore, RMS, PSD, and standard deviation (named y1, y2, and y3, respectively) are considered for classification of states. The range of features and their weights to identify normal, preseizure, and seizure states are determined. The features are selected using ANOVA feature selection, as discussed earlier.

4.5.1 Feature selection ANOVA

Univariate ANOVA is a statistical tool used to analyze the means of different datasets to test variation between two or more means. It is a very useful tool for feature analysis and classification. The technique can be used for the nonspecific NULL hypothesis, feature selection, classification, and prediction. We consider the classification of epileptic EEG signals in preseizure, seizure, and normal states. Features extracted for different patients include mean, variance, and RMS. Let us consider the RMS feature for

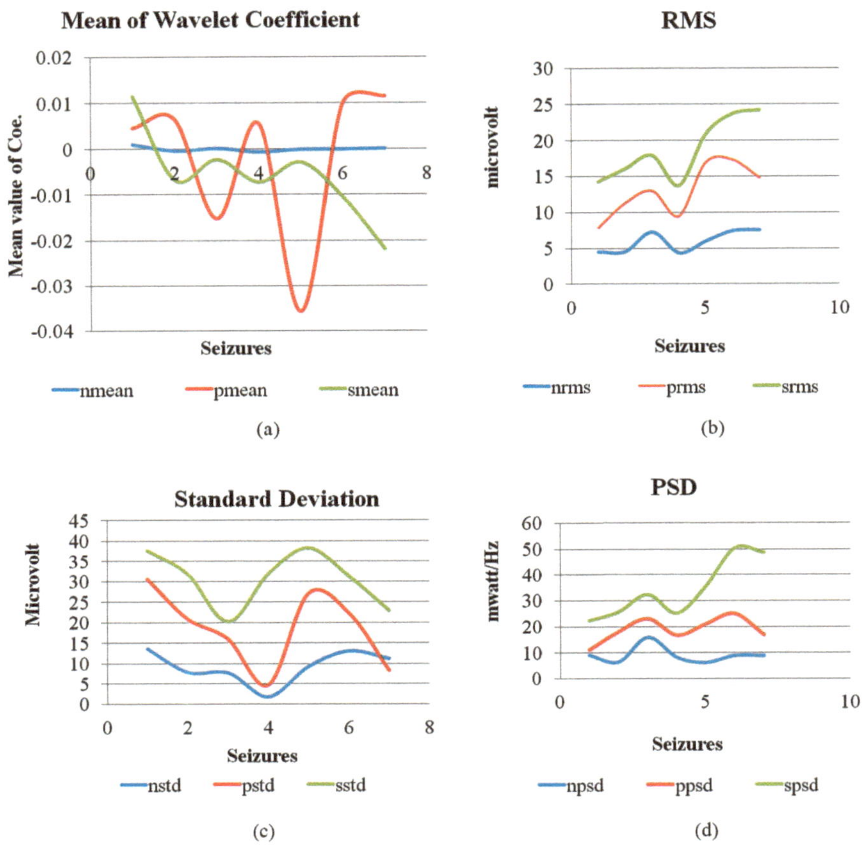

Figure 4.6 Features using pattern adapted wavelet transform during the normal, preseizure, and seizure states for seven EEG signals containing a single seizure each. (a) Mean. (b) Root mean square (RMS). (c) Standard deviation. (d) Power spectral density (PSD).

preseizure (Prms), seizure (Srms), and normal (Nrms) states. ANOVA is used to identity the usefulness of the RMS in this research by comparing Prms and Srms with Nrms. Table 4.3 shows Srms, Prms, and Nrms values obtained with a window size of 256 samples, and Table 4.4 shows ANOVA test parameters for checking the importance of the RMS feature. Table 4.4 justifies the calculations performed during ANOVA and its relevance using the *P*-value and the F-measure. If $P < 0.05$ and the F-measure is larger than F critical, the NULL hypothesis is rejected.

Table 4.5 shows *P*-values, where parameters having $P < 0.05$ are relevant for classification. As shown in Table 4.6 the ANOVA parameters for

Table 4.4 Features extracted using pattern adapted wavelet transform during normal (N), preseizure (PS), and seizure (S) states continued.

Feature vector	RMS N	RMS PS	RMS S	STD N	STD PS	STD S	PSD N	PSD PS	PSD S
F(2,1,Kavg)	4.52	7.83	14.23	13.71	730.47	37.56	9.15	11.02	22.39
F(2,2,Kavg)	4.55	11.26	16.02	7.99	20.84	31.75	6.49	17.85	25.66
F(2,3,Kavg)	7.28	12.95	17.90	7.79	15.86	20.24	15.92	23.00	32.30
F(2,4,Kavg)	4.35	9.46	13.66	1.82	4.62	31.68	8.33	16.58	25.20
F(2,5,Kavg)	5.99	16.89	20.84	9.36	26.95	38.15	6.25	20.86	35.35
F(2,6,Kavg)	7.45	17.33	23.71	13.14	22.26	31.23	8.94	24.98	50.34
F(2,7,Kavg)	7.58	14.79	24.23	11.31	8.26	22.90	8.87	16.76	48.73

Table 4.5 RMS values for seizure (Srms), preseizure (Prms), and normal (Nrms) states.

Seizure number	RMS N	RMS PS	RMS S
1	93.1325	48.9329	41.9373
2	106.135	40.6979	32.2911
3	103.029	70.1816	62.0762
4	109.398	82.8107	42.8806
5	100.52	85.5259	25.3026
6	59.9825	56.242	46.1823

Table 4.6 ANOVA: Single factor test showing the importance of the RMS feature.

Anova:						
Group	*Count*	*Sum*	*Average*	*Variance*		
Srms	6	572.19	95.36	331.02		
Prms	6	384.39	64.06	337.11		
Nrms	6	250.67	41.77	158.82		
Source	*SS*	*Df*	*MS*	*F*	*P-value*	*F crit*
Between groups	86.2	2	4348.1	15.77	0.000205*	3.68
Within groups	4134.81	15	275.654			

RMS feature are calculated and it specifies that the $P < 0.05$ for RMS feature thus it is useful for classification. By observing P-values of all the features mentioned in Table 4.7, the variance, RMS, kurtosis, Sum, and power can be used to construct the feature vectors.

Table 4.5 shows the P-values calculated for different seizure file features to check their relevance. It is observed that Patient 6 files have 10 seizures, thus showing maximum relevance for accurate classification. The feature's variance, RMS, kurtosis, sum, power, and ZCR have $P < 0.05$. It means that these features have less similarity in the three categories of seizure, preseizure, and normal. Thus the NULL hypothesis is rejected.

Based on feature selection performed for time and frequency and pattern adapted wavelet features, the feature vectors f1 and f2 are constructed. These are inputs to the fuzzy model to classify the EEG signals as preseizure, seizure, and normal. For training preseizure characteristics, each signal in the training set is segmented as preseizure, seizure, and normal. The relevant features are calculated, and the threshold for each category is determined to construct fuzzy rules for the fuzzy classifier. Fig. 4.7 shows the final vectors to be given to the fuzzy classifier. The inputs for the classification are:

Table 4.7 *P*-value stating its similarity index by hypothesis testing.

Patient	Seizures	Mean	VAR	COV	RMS	Kurt	Sum	Power	ZCR
P1	6	0.44	0.0001	0.21	0.0002	0.36	0.0002	0.0001	0.9
P2	3	0.69	0.98	0.29	0.99	0.09	0.07	0.98	0.22
P3	7	0.228	0.0002	0.46	0	0.0027	0.0001	0.0002	0.25
P4	4	0.13	0.947	0.54	0.89	0.028	0.0096	0.94	0.36
P6	10	0.08	0.04	0.47	0.058	0.03	0	0.04	0.0049
P7	3	0.54	0.47	0.64	0.50	0.09	0.0722	0.47	0.51

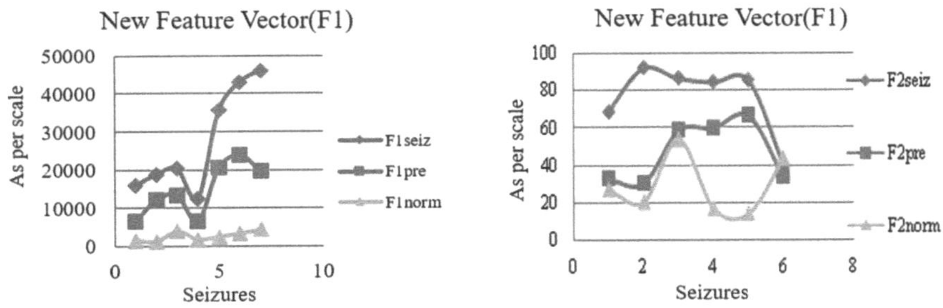

Figure 4.7 Feature vectors given as input to the fuzzy classifier. (a) Feature vector from time and frequency domains. (b) Feature vectors from wavelet features.

1. $f1 = a1x1 + a2x2 + a3x3 + a4x4 + a6x6$ (note that x5 [COV] is neglected),
2. $f2 = b2y2 + b3y3 + b4y4$ (note that y1 [mean] is neglected).

The coefficients are calculated on the basis of percentage variation of seizure features with reference to the feature values of the normal state. The threshold of each class was determined by analyzing maximum and minimum values of the features. Based on variation between the three states, weights a1–a6 and b1–b4 are selected, and fuzzy rules are constructed. For each selected feature, its percentage variation is observed and its trend line with reference to its value during the seizure state and the weight of each coefficient is constructed. The constructed feature vectors are shown in Fig. 4.7. Distinct values are observed in the three classes, and thus, the fuzzy classifier can categorize them into seizure, preseizure, and normal states.

The fuzzy model with f1 and f2 input to classify signals into the states of the brain is used with outputs normal, preseizure, and seizure states.

The rules used to classify the brain states are:

1. If f1 is low and f2 is low, then output is "normal state."
2. If f1 is low and f2 is moderate or f1 is moderate and f2 is moderate, then output is "preseizure state."
3. If f1 is high and f2 is high or f1 is moderate and f2 is high, then output is "seizure state."

The fuzzy-based classification procedure used in this method is given as follows:

- Construct two feature vectors. The f1 feature vector is constructed using time domain features and frequency domain features and the f2 feature vector is derived from features calculated from PAWT.

Table 4.8 Performance metrics of the system.

Total number of patients	Number of total events	Actual number of seizures	TP	TN	FP	FN
23	352	129	125	205	12	10

Table 4.9 Overall performance of preseizure detection.

TPR (%)	FPR	Specificity (%)	Sensitivity (%)	Accuracy (%)
96.89	0.0552	94.47	92.59	93.75

- Triangular membership functions are used for input and output of the fuzzy system. These are calculated from minima, averages, and maxima of feature vectors and rated as "low," "moderate," and "high," respectively.
- These new feature vectors are calculated and based on percentage variation; the threshold is determined to distinguish between the preseizure and seizure states. Two feature vectors, f1 and f2, are inputs to the fuzzy model.
- If f1 is low and f2 is low, then output is "normal state."
- If f1 is low and f2 is moderate or f1 is moderate and f2 is moderate, then output is "preseizure state."
- If f1 is high and f2 is high or f1 is moderate and f2 is high, then output is "seizure state."

4.6 Performance evaluation

Considering the spatial and temporal context of channels, features are analyzed and classified using the fuzzy classifier. Performance metrics include true positive (TP), false positive (FP), true negative (TN), and false negative (FN), as shown in Table 4.8. The performance of the system is evaluated by these metrics, and the true positive rate (TPR) and false positive rate (FPR) are calculated, as shown in Table 4.9. Overall system implementation is verified based on classification accuracy, sensitivity, and specificity. The TPR is the probability of identifying epileptic seizures correctly, and it is also termed sensitivity. It is calculated as follows:

$$TPR = \frac{Total\ number\ of\ TPs}{Total\ number\ of\ exisiting\ seizures} \times 100. \tag{4.3}$$

The FPR is the number of false positives detected in nonseizure EEG signals. It is also related to specificity. Specificity is the fraction of incorrectly

Table 4.10 Performance comparison with existing work.

Method	**Mohamed Shakir et al. [15] fuzzy classifier**	**Seizure detection**	**Preseizure detection**
TPR	95%	97.67%	96.89%
FPR	5%	3.52%	3.52%
Specificity	93%	95.34%	94.50%
Accuracy	93.47%	95.94%	94.75%

mapped nonseizures of the total number of errors. Specificity and FPR provide the same information but are not the same. FPR is calculated as follows:

$$FPR = \frac{Total\ Number\ of\ FPs}{Total\ number\ of\ exisiting\ seizures}. \tag{4.4}$$

Classification accuracy is the ratio of the number of correctly classified frames into seizures and nonseizures to the overall detected seizures:

$$Accuracy = \frac{TP + TN}{TP + TN + FP + FN}. \tag{4.5}$$

Sensitivity is the rate of correctly detected seizures against the actual number of seizures detected by the system.

$$Sensitivity = \frac{TP}{TP + FN}. \tag{4.6}$$

Considering seizure detection, the specificity is the fraction of corectly classifying normal subjects against total normal subjects. For detecting non-epileptic patients from total seizure patients the specificity is the fraction of all correctly classified nonseizures against the total number of non-epileptic patients:

$$Specificity = \frac{TN}{TN + FP}. \tag{4.7}$$

During experimentation, the data of 22 patients from the database are used. The first 6 patients are used for training, 4 patients are used for validation, and 12 patients are used for testing. The major objective of the work is to detect preseizure time and thus to predict seizures. The results are compared with an existing prediction system in Table 4.10. The accuracy and FPR are improved. Table 4.11 shows results obtained for preseizure detection. As shown, the maximum time before the proposed algorithm

Table 4.11 Preseizure detection for patient 1.

Files	Seizure starts (sec)	Seizure ends (sec)	Window index start	Window index stop	Pre-seizure time	Early detection time
chb01_03	2996	3036	1498	1518	2971	25
chb01_04	1467	1494	734	747	1422	45
chb01_15	1732	1772	866	886	1682	50
chb01_16	1015	1066	508	533	995	20
chb01_18	1720	1810	860	905	1679	41
chb01_21	327	420	164	210	315	12
chb01_26	1862	1963	931	982	1794	68

can detect the occurrence of seizures is about 110 seconds. With this observation, the accuracy of preseizure detection is observed to be the same as that of seizure detection, but the FPR is also as high as that for seizure detection.

4.6.1 Preseizure detection for seizure prediction

Quantitative EEG analysis is an essential tool to extract timing parameters and frequency components. A suitable method for analyzing highly dynamic, nonlinear, and nonstationary EEG signals entails the use of flexible and nonsubjective measures. The database is tested for preseizure detection and its time duration, as shown in Table 4.11.

4.7 Summary

This implemented system is trained for various time and frequency domain features, and it is observed that EEG signals of various patients have characteristics of increased amplitude and frequency variation. The signals were analyzed with features and a fuzzy classifier model to improve performance, and the presented method could achieve a classification accuracy of 92.17%, a TPR of 96.89%, and an FPR of 34.8%. The fuzzy logic classifier model is used to classify transitions between brain states. The preseizure time observed in each case is different, so on average seizures are predicted approximately 30 sec before occurrence, and thus the accuracy is improved with the implemented fuzzy inference system.

References

[1] Y. Tang, D.M. Durand, A tunable support vector machine assembly classifier for epileptic seizure detection, Expert Systems with Applications 39 (4) (2012) 3925–3938.

[2] B. Karlik, S.B. Hayta, Comparison machine learning algorithms for recognition of epileptic seizures in EEG, in: Proceeding of International Work-Conference on Bioinformatics and Biomedical Engineering, Granada, 2014, pp. 1–12.

[3] N. Sadati, H.R. Mohseni, A. Maghsoudi, Epileptic seizure detection using neural fuzzy networks, in: Proceeding of IEEE International Conference on Fuzzy Systems, Vancouver, BC, 2006, pp. 596–600.

[4] U. Qidwai, A.S. Malik, M. Shakir, N. Kamel, Hardware simulator for seizure, preseizure and normal mode signal generation in LabVIEW environment for Research, International Journal of Bioscience, Biochemistry and Bioinformatics 3 (6) (2013) 588–592.

[5] B. Atoufi, C. Lucas, Ali Zakerolhosseini, A survey of multi-channel prediction of EEG signal in different EEG states: normal pre-seizure, and seizure, in: Proceeding of Seventh International Conference on Computer Science and Information Technologies, 2009, pp. 327–331.

[6] P. Khosropanah, A.R. Ramli, R. Ashurov, A. Ahmedov, Detection of epileptic spikes in egg signal using wavelet transform and adaptive neuro, in: Fuzzy Inference System (Anfis) Technique, Proceeding of International Conference on Agricultural and Food Engineering for Life (Cafei2012), 2012, pp. 583–590.

[7] D. Najumnissa, T.R. Rangaswamy, Detection and classification of epileptic seizures using wavelet feature extraction and adaptive neuro-fuzzy inference system, International Journal Of Computational Engineering Research (IJCER) 2 (3) (2012) 755–761.

[8] S.-H. Oh, Yu-Ri Lee, H.-N. Kim, A novel EEG feature extraction method using Hjorth parameter, International Journal of Electronics and Electrical Engineering 2 (2) (2014) 106–110.

[9] A. Subasi, Automatic detection of epileptic seizure using dynamic fuzzy neural networks, Expert Systems with Applications 31 (2) (2006) 320–328.

[10] I. Guler, E.D. Ubeyli, Adaptive neuro-fuzzy inference system for classification of EEG signals using wavelet coefficients, Journal of Neuroscience Methods 148 (2) (2005) 113–121.

[11] A.F. Rabbi, A. Aarabi, R. Fazel-Rezai, Fuzzy rule-based seizure prediction based on correlation dimension changes in intracranial EEG, in: Proceeding of Annual International Conference of the IEEE Engineering in Medicine and Biology, 2010, pp. 3301–3304.

[12] D. Gajic, Z. Djurovic, S. Di Gennaro, Fredrik Gustafsson, Classification of EEG signals for detection of epileptic seizures based on wavelets and statistical pattern recognition, Biomedical Engineering: Applications, Basis and Communications 26 (2) (2014).

[13] P. Jaiswal, R. Koushal, EEG signal classification using modified fuzzy clustering algorithm, International Journal of Computer Science and Information Technologies 6 (3) (2015) 2031–2034.

[14] U. Qidwai, A.S. Malik, M. Shakir, Embedded fuzzy classifier for detection and classification of preseizure state using real EEG data, in: Proceeding of 15th International Conference on Biomedical Engineering, IFMBE, 43, 2014, pp. 411–415.

[15] M. Shakir, A.S. Malik, N. Kamel, U. Qidwai, Intelligent fuzzy classifier for pre-seizure detection from real epileptic data, in: Proceeding of Science and Information Conference (SAI), 2014, pp. 276–279.

[16] M. Jothi, N.B. Balamurugan, R. Harikumar, Design and implementation of VLSI fuzzy classifier for biomedical application, International Journal of Innovative Research in Science, Engineering and Technology 3 (3) (2014).

[17] O. Salem, A. Naseem, A. Mehaoua, Epileptic seizure detection from EEG signal using discrete wavelet transform and ant colony classifier, in: Proceeding of IEEE International Conference on Communications (ICC), 2014, pp. 3529–3534.
[18] B. Pushpa, D. Najumnissa, Classification of epileptic seizure EEG signals using EMD and ANFIS, in: Proceeding of International Conference on Informatics, Electronics and Vision (ICIEV), 2014, pp. 1–5.
[19] J. Wang, X.Z. Gao, J.M.A. Tanskanen, P. Guo, Epileptic EEG signal classification with ANFIS based on harmony search method, in: Proceeding of Eighth International Conference on Computational Intelligence and Security (CIS), 2012, pp. 690–694.
[20] M. Sood, S.V. Bhooshan, Automatic processing of EEG signals for seizure detection using soft computing techniques, in: Recent Advances and Innovations in Engineering (ICRAIE), 2014, pp. 1–6.
[21] H. Niknazar, K. Maghooli, A.M. Nasrabadi, Epileptic seizure prediction using statistical behavior of local extrema and fuzzy logic system, International Journal of Computer Applications 113 (2) (2015) 24–30.
[22] C.J. James, R.D. Jones, P.J. Bones, G.J. Carroll, Spatial analysis of multi-channel EEG recordings through a fuzzy-rule based system in the detection of epileptiform events, in: Proceedings of the 20th Annual International Conference of the IEEE on Engineering in Medicine and Biology Society, 1998, pp. 2175–2178.
[23] A.K. Sinha, K.A. Loparo, W.J. Richoux, A new system theoretic classifier for detection and prediction of epileptic seizures, in: Proceeding of 26th Annual International Conference of the IEEE on Engineering in Medicine and Biology Society (IEMBS'04), 2004, pp. 415–418.
[24] K. Gopika Gopan, A. Harsha, L.A. Joseph, E.S. Kollialil, Adaptive neuro-fuzzy classifier for 'Petit Mal' epilepsy detection using mean teager energy, in: International Conference on Advances in Computing, Communications and Informatics (ICACCI), Mysore, 2013, pp. 752–757.
[25] R. Sukanesh, R. Harikumar, A structured soft (max-min) decision trees for patient specific fuzzy classifier in the classification of epilepsy risk levels from EEG signals, in: International Conference on Computational Intelligence and Multimedia Applications, Sivakasi, Tamil Nadu, 2007, pp. 435–442.
[26] A.L. Goldberger, L.A. Amaral, L. Glass, J.M. Hausdorff, P.C. Ivanov, R.G. Mark, J.E. Mietus, G.B. Moody, C.K. Peng, H.E. Stanley, PhysioBank, PhysioToolkit, and PhysioNet: components of a new research resource for complex physiologic signals, Circulation 101 (23) (2000) E215–E220, https://doi.org/10.1161/01.cir.101.23.e215, PMID: 10851218.
[27] V.K. Harpale, V.K. Bairagi, Time and frequency domain analysis of EEG signals for seizure detection, in: Proceeding of the IEEE International Conference on Microelectronics, Communication & Computing (MicroCom 2016), NIT, Durgapur, 2016.
[28] Hector Mesa, Adapted wavelets for pattern detection, in: Progress in Pattern Recognition, Image Analysis and Applications, in: Lecture Notes in Computer Science, vol. 3773, 2005, pp. 933–944.
[29] V.K. Harpale, V.K. Bairagi, An adaptive method for feature selection and extraction for classification of epileptic EEG signal insignificant states, Journal of King Saud University: Computer and Information Sciences (ISSN 1319-1578) (2018), https://doi.org/10.1016/j.jksuci.2018.04.014.
[30] V. Harpale, V. Bairagi, FPGA based architecture implementation for epileptic seizure detection using one way ANOVA and genetic algorithm, Biomedical and Pharmacology Journal 12 (3) (2019) 1543–1553.

CHAPTER 5

Effective method for epileptic and nonepileptic seizure classification

5.1 Introduction

Abnormal movement and unconsciousness are responsible for the generation of seizures, which in turns cause temporary loss of control. Seizures observed due to sudden abnormal electrical discharges in the brain are called epileptic seizures. Stress-related seizures are referred to as psychogenic nonepileptic seizures (PNES). Such nonepileptic seizures (NES) are similar to epileptic seizures and are hence often misdiagnosed as epileptic seizures. Approximately 70–80% of cases are observed in women. PNES are NES caused by stress, not by electrical discharge as in epilepsy, and are thus referred to as pseudoseizures. They should not be treated with antiepileptic drugs (AEDs); therefore, there is a requirement of the classification of PNES and epileptic seizures.

Video-electroencephalography (V-EEG) is a preferable tool for the classification of nonepileptic and epileptic seizures, but in rural areas, this tool is observed to be very costly and there is lack of information about the tool and its handling, so there is a need for simple instrumentation. Scalp EEG can be a simple solution to analyze the status of patients in rural areas. EEG is a noninvasive tool, which can provide information at a high temporal resolution to assess dynamic brain activity. If risk analysis is also offered, then doctors in rural areas could provide proper medication and avoid death in severe cases.

5.2 Dataset used for experimentation

The previous experimentation was conducted based on the CHB-MIT database [1] with 23 subjects and 10 subjects database (five healthy participants and five epileptic patients) from Andrezejak (2001). For this work, the TUH EEG Seizure Corpus (TUSZ) (American Clinical Neurophysiology Society in Phoenix, Arizona, USA) dataset [2] is used. As discussed earlier, the dataset is divided into a training set and a testing set, with 264 subjects and 50 subjects, respectively. In the training set, 1989 files are

Brain Seizure Detection and Classification Using EEG Signals
https://doi.org/10.1016/B978-0-32-391120-7.00010-4

Seizure Type	Seizure Class	Description:
FNSZ	Focal Non-Epileptic Seizure	Unspecified Non-epileptic Focal seizures
GNSZ	Generalized Non-Epileptic Seizure	Unspecified Non-epileptic Generalized seizures
SPSZ	Simple Partial Seizure	Partial seizures during consciousness;
CPSZ	Complex Partial Seizure	Partial Seizures during unconsciousness;
ABSZ	Absence Seizure	Loss of consciousness for a few seconds
TNSZ	Tonic Seizure	Stiffening of the body during a seizure
CNSZ	Clonic Seizure	Jerking/shivering of the body during a seizure
TCSZ	Tonic-Clonic Seizure	At first stiffening and then jerking of a body
ATSZ	Atonic Seizure	Sudden loss of muscle tone
MYSZ	Myoclonic Seizure	Myoclonous jerks of limbs
NE	Non-Epileptic	Non-Epileptic Patients without seizures
Norm	Normal	Patients not affected by Neurological Disorders

Figure 5.1 Seizure classes used from the TUH database.

handled for experimentation, with 1254 seizure files, 145 non-epileptic seizure files and 590 normal files, and in the testing set, 1015 files are handled, with 679 seizure files, 89 non-epileptic seizure files and 247 normal files, in European Data Format (EDF). The subcategories of seizures considered for classification are focal NES (FNSZ), focal NES (GNSZ), simple partial seizure (SPSZ), complex partial seizure (CPSZ), absence seizure (ABSZ), tonic seizure (TNSZ), clonic seizure (CNSZ), tonic-clonic seizure (TCSZ), atonic seizure (ATSZ), and myoclonic seizure (MYSZ), along with nonepileptic without seizures and normal. A detailed description of classes with their effects on the body during the seizures is shown in Fig. 5.1.

Seizure files in the TUSZ database were recorded in EDF file format with various methods, such as an epilepsy monitoring unit (EMU) with 32 channels, an intensive care unit (ICU), cardiac intensive care, Neuro-ICU, neural surgical ICU, pediatric ICU, etc. Table 5.1 describes the event files

Table 5.1 Seizure events from the TUH database.

Seizure type	Training event files	Testing event files
FNSZ	663	284
GNSZ	129	187
SPSZ	48	6
CPSZ	320	92
ABSZ	52	42
TNSZ	17	50
TCSZ	25	17
MYSZ	0	1
Nonepileptic	145	89
Normal	590	247
Total files	1989	1015

Table 5.2 Patient information and remarks.

Patient ID	Treatment type	Seizure frequency	Type of seizure	Remark
2289	EMU	No Seizures	–	Nonepileptic other brain disorder
4671	ICU	39	GNSZ	Nonepileptic
8460	ICU	10 (long duration)	CPSZ	Epileptic AED application

used for experimentation in the training and testing sets. The patients or subjects, methods of EEG signal recording, seizure frequencies, and types of seizure are labeled in the database with patient identification number, as shown in Table 5.2.

5.2.1 Mathematical model of EEG signals

The mathematical model states the static and dynamic behavior of the EEG signal in terms of various parameters. The EEG signal is a continuous-time signal. Temporal characteristics of EEG signals over locations are identified by various channels for continuous time "t." The EEG signal over the continuous time t, if sampled with a sampling time T, is modeled as

X[n] = X(nT), n = 1, 2, 3, $T = 1/S$, where "S" is the sampling rate. Here X[n] is the discrete time signal sampled at time intervals t = T, 2T,……., nT.

- f(EEG)={N, x[n], A, R, S}
- $y[n]= [A(x_m[n]+n_e [n])@R]_{N \times n}$

Where N= no. of channels

A= Pre-processing Amplification factor

$x_m[n]$ = EEG signal of mth channel

$= \sum EPSP + \sum IPSP$ m=1: N

R= bit resolution of ADC

S=Sampling Rate

n: sample instant =1: t*S

t= recording time in seconds

Excitatory post-synaptic potential (EPSP)

Inhibitory post-synaptic potential (IPSP)

Figure 5.2 Methodology of seizure classification.

The behavior of the signal over the total time is called the dynamics of signal y[n]. This state signifies $y[n] \in \Re^N$, where $\Re^N$ is the phase space of the system. X[n] is a discrete time measured from the single channel behavior of the brain.

Mathematically, the resultant behavior is a summation of the true signal and added noise to it:

$y[n] = X(nT) + \eta\ (nT),\ n = 1, 2, 3, \ldots,$

where X (nT) is the value of the true signal at time t = nT (where T is the sampling interval) and η (nT) is the error of this measurement at this time. If η (nT) is small, then $y(n) \approx X[nT]$.

More specifically, considering all parameters, the mathematical model for multichannel EEG can be defined as shown in Fig. 5.2.

5.3 Seizure classification

The primary objective of the chapter is to classify EEG signals for identification of seizures and its classes as mentioned above. It mainly focuses on the classification of epileptic and nonepileptic seizures using existing methods and the new effective method with their performance comparison.

A NES is a random behavior that resembles seizure behaviors but is not caused by abnormal electrical discharges in the brain. The leading causes of NES are low blood pressure or body fluid or chemical imbalance, and are typically caused by psychological issues such as handling thoughts, memories, emotions, or sensations in the brain. Such problems are sometimes related to stress. PNES is observed in approximately 10–20% of seizure

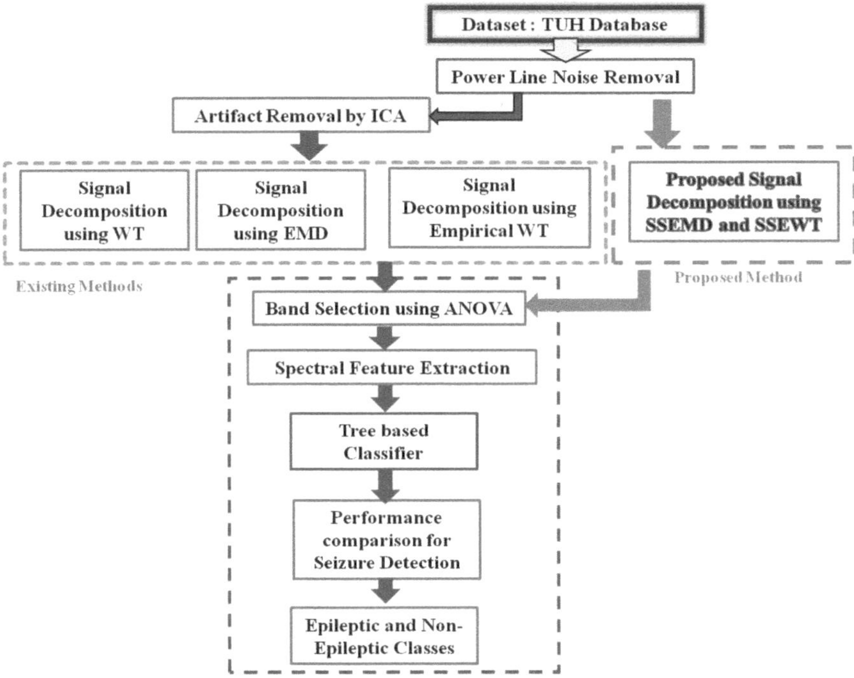

Figure 5.3 Methodology of seizure classification.

patients; in other words, one of every five to ten seizure patients sent to epilepsy centers is found to have NES but is diagnosed as epileptic. The NES features such as frequency, triggers, and duration of seizures vary from epileptic features. The term "nonepileptic seizure" (NES) covers both physiologically and psychologically mediated paroxysmal events similar to epileptic manifestations. The psychogenic term is added to NES to indicate psychologically mediated nonepileptic events. The experimentation design flow discussed here is shown in Fig. 5.3, which describes the overall methodology. The EEG signals recorded in the database have noise and artifacts caused by power noise, eye blinking, chewing, etc. The effect of powerline noise of 60 Hz is observed with all 32 channels of the recorded EEG signal. The notch FIR filter is designed with a center frequency of 60 Hz with order 18 for efficient removal.

Fig. 5.4 shows noise in EEG channel 10 along with the filter response and the result of order 10 and 18. It is observed that an order of 18 results in good performance. Thus, each EEG signal is processed with the EEGLAB

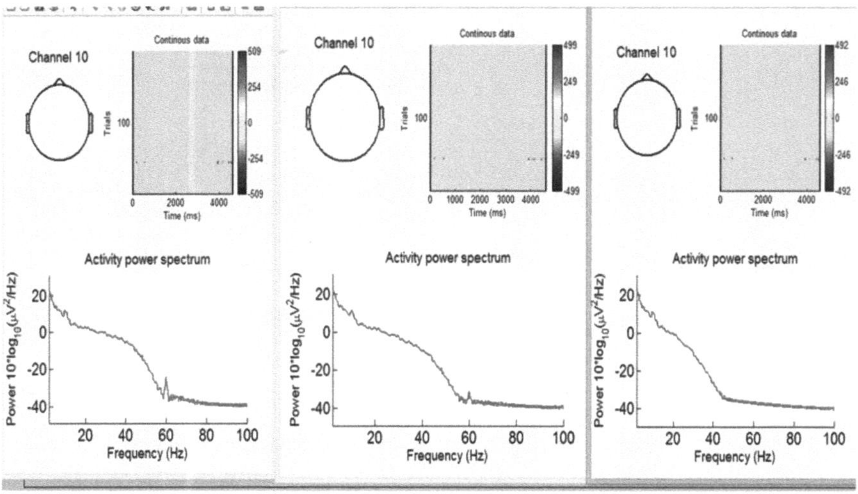

Figure 5.4 FIR filter response for order of 10, 15, and 18.

toolbox of MATLAB® to remove this powerline noise and restore the signal in EDF format.

Firstly, the features are extracted by a well-established feature extraction method for nonstationary signals such as wavelet transform (WT), empirical mode decomposition (EMD), or empirical WT (EWT). The appropriate methods are selected based on analysis of variance (ANOVA), and events are trained and classified with a random forest classifier.

Secondly, the procedure is repeated for SSEMD and SSEWT feature extraction methods, and the results are compared.

5.3.1 Feature extraction using DWT, EMD, and EWT

EEG signals are recorded from various electrodes placed with the 10–20 electrode placement standard. The EEG signal used has 32 EEG channels, and every channel has an impact on artifacts. The artifacts are removed using independent component analysis (ICA), and the signals are used further for feature extraction and classification. As discussed in EEG pre-processing, the EEGLAB toolbox of MATLAB is used to remove artifacts using ICA, and after artifact removal the preprocessed EEG signal is saved as EDF files. The preprocessed database is then used for seizure classification.

The features extracted by WT, EMD, and EWT are power spectral density (PSD), zero crossing rate (ZCR), kurtosis, spectral entropy, power, spectral centroid (SC), spectral slope (SS), high-frequency content (HFC),

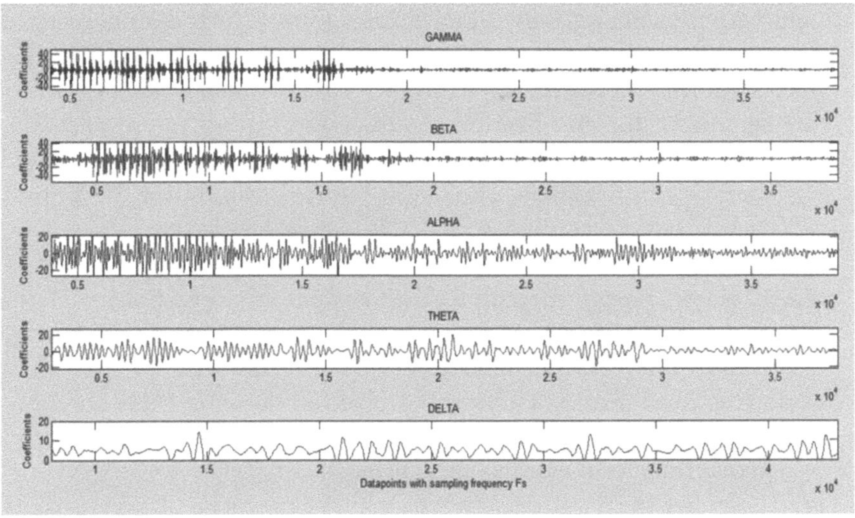

Figure 5.5 Wavelet decomposition of five subbands with seizure activity.

spectral crest measure (SCM), spectral roll-off, the spectral flatness measure (SFM), spectral flux (SF), the upper bound (ub), and the lower bound (lb).

After hypothesis testing for finding relevant features with reference to seizure activity, the selected features are PSD, spectral entropy, ZCR, and SF. These features are submitted to the classifier for effective classification of seizures.

1. **Feature extraction by wavelet transform**

   For the multifrequency EEG signal, the discrete WT (DWT) is observed to be a useful selection for analysis as compared to continuous WT. As discussed earlier, the EEG signal is decomposed using a Daubechies wavelet function (DB4) with four levels of decomposition or five subbands called delta (0.1 to 3.5 Hz), theta (4 to 7.5 Hz), alpha (8 to 13 Hz), beta (14 to 30 Hz), and gamma (> 30 Hz) bands, as shown in Fig. 5.5.

   DWT filter decomposition gives separate frequency bands corresponding to various brain activities, and the significance of the decomposed signal is tested by applying ANOVA. It is observed that for seizure detection, the gamma band and the beta band are more relevant. The SS, SF, centroid, roll-off, crest, entropy, and tone power ratio are extracted from the gamma and beta frequency bands. So some of the spectral features for each seizure subcategory are extracted from the gamma and beta frequency bands, and a further feature vector is constructed for classification. It is visualized that these spectral features show differentiation in values in the seizure subcategory, and thus, all features are significant for classification. As discussed in the feature selection, the most relevant features are selected after comparing all the extracted features for all the classes. Thus most relevant features for the gamma and beta bands are entropy, PSD, ZCR, SF, and SS.

2. **Feature extraction by empirical mode decomposition**
EMD is an adaptive method for the analysis of nonstationary physiological signals. The intrinsic mode functions (IMFs) are selected by calculating one zero between successive extrema, and an IMF must have zero local means. The steps involved in deriving IMFs are given as follows:
   - Extract the local extrema of signal x(t).
   - Calculate the envelope using the cubic spline interpolation method.
   - Calculate the mean of the envelope me(t).
   - Calculate Di(t)=x(t)-me(t).
   - If mean of Di(t) < Threshold, then IMF=Di(t); else repeat the procedure for Di(t)
   - Next the IMF will be derived by x1(t)= x(t)-IMF; repeat the full procedure for x1(t) until the stopping condition is satisfied.

   For experimentation here, five IMFs are calculated, the relevant IMFs are selected, and features are extracted for classification. Similar to the previous feature selection discussion, all IMFs are tested using ANOVA, and only IMF1 and IMF2 are used for further feature extraction. The features are extracted for all the classes discussed earlier, and only relevant features are considered for classification.

3. **Feature extraction by empirical wavelet transform**
The EMD method has many limitations; for example, it does not use any predefined basis function, but it is derived on the basis of data, and thus lacks mathematical theory. The IMF selection procedure is not optimized, it suffers from a mode mixing problem, and its computational efficiency is very low. Due to the ad hoc approach of EMD, it is difficult to model the EMD function, so for stabilized solutions [3], ensemble EMD (EEMD) was proposed, but it has high computational cost due to averaging each of the EMD components. A new approach, EWT, was proposed by J. Gilles in 2013 to solve the mode mixing problem of EMD [4]. Compared with EEMD

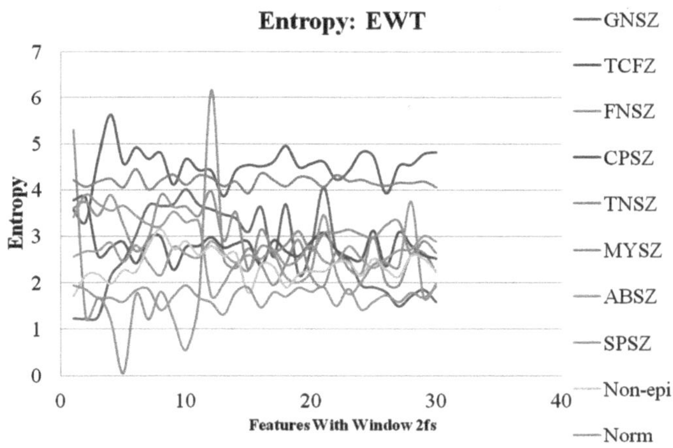

Figure 5.6 EWT-based extracted feature: entropy.

and EMD, EWT has improved performance in terms of computation time, mode estimates, and robustness to noise [5]. EWT uses concepts of EMD and wavelet decomposition, and the appropriate wavelet filter bank extracts intrinsic modes. In EMD, the Fourier spectrum is segmented for IMF decomposition. The detailed and approximation coefficients are calculated using the wavelet concept as stated by

$$IMF_1 = W_x(1, t) = IFFT(X(\omega) * \psi_1(\omega)), \tag{5.1}$$

$$IMF_N = W_x(N, t) = IFFT(X(\omega) * \psi_N(\omega)), \tag{5.2}$$

where "N" is the number of simulated IMF components, $N = 1, \ldots, n-1$, and $\psi_n(\omega)$ is the filter constructed using wavelets.

The sensitive IMF is selected based on the probability of the F-measure calculated using ANOVA for healthy and unhealthy EEG signals. The features are extracted, and a feature vector is constructed for classification of epileptic and normal EEG signals. The relevant features SF, SS, PSD, ZCR, standard deviation, and entropy are extracted from IMF1 and IMF2, as shown in Figs. 5.6–5.10. These features have different values for each of the targeted class, and arethus useful for classification. Fig. 5.6 shows entropy measures for all the classes enlisted. The values overlap; thus, the weight of the entropy feature is less compared to other features when performing feature selection. Fig. 5.7 shows PSD

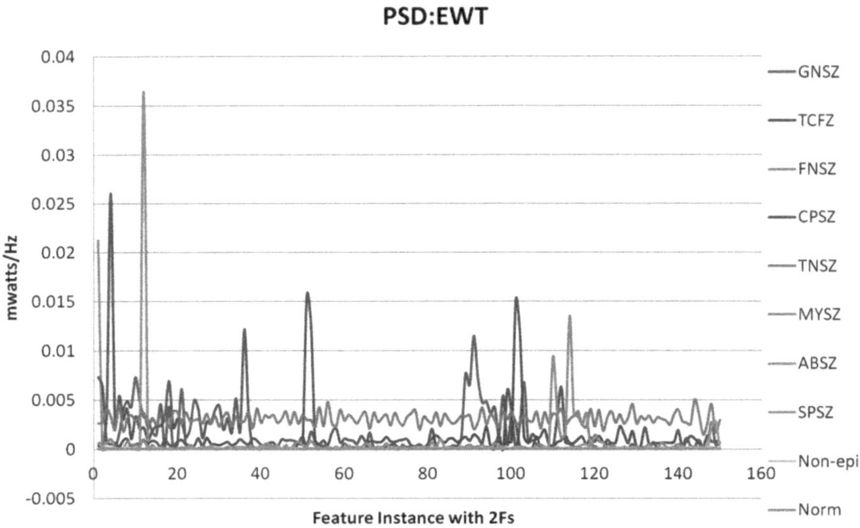

Figure 5.7 EWT-based extracted feature: power spectral density (PSD).

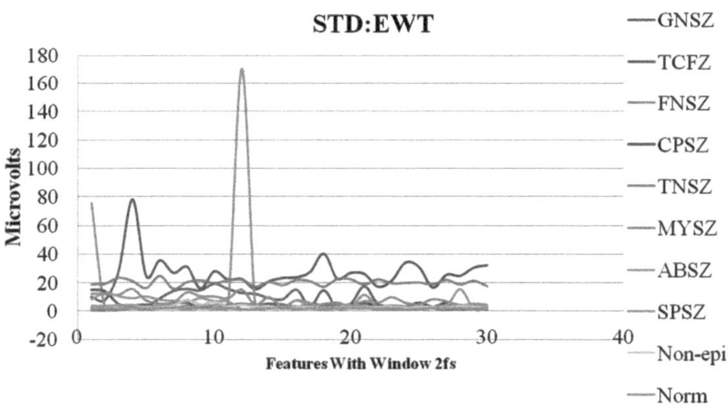

Figure 5.8 EWT-based extracted feature: standard deviation (STD).

feature values for all classes, but its relevance is low as compared to spectral features.

Fig. 5.8 shows standard deviation values. It shows this feature is useful for classification of normal, GNSZ, and nonepileptic classes. Figs. 5.9 and 5.10 show spectral features, i.e., SS and SF, respectively, that are most relevant for classification due to their distinctive values in different classes.

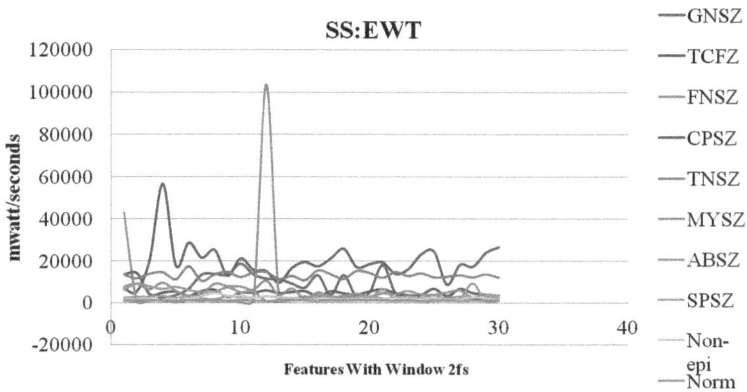

Figure 5.9 EWT-based extracted feature: spectral slope (SS).

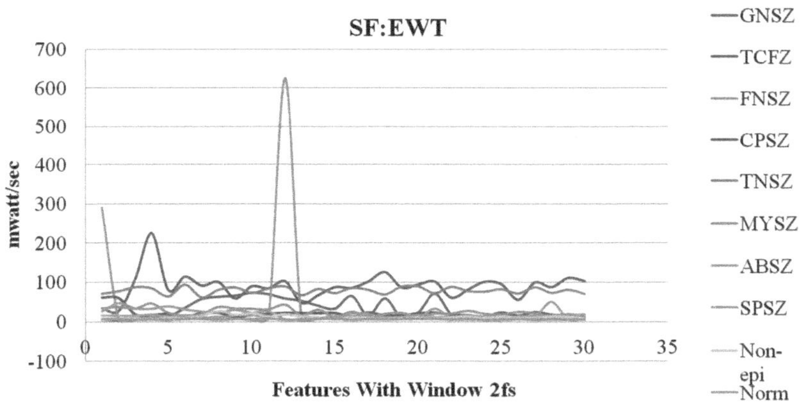

Figure 5.10 EWT-based extracted feature: spectral flux (SF).

5.3.2 Effective method for classification of seizures

The EMD and EWT methods discussed in the previous section have limitations; for example, EMD suffers from a mode mixing problem and EWT needs fine spectral segmentation to achieve good performance of the system. These problems are addressed by the proposed singular spectrum EMD (SSEMD) and singular spectrum EWT (SSEWT). These methods improve the classification by significantly improving accuracy and reducing the number of false positives. SSEWT is superior amongst all the methods discussed here with respect to mode selection, boundary selection, and noise removal. The adaptive boundaries are derived in EWT and filtered by wavelet decomposition, but the IMF boundaries overlap, which reduces the performance of EWT. The method proposed here provides a solution to this overlapping problem of IMF boundaries.

Most of the time, ICA is used to remove the unwanted artifacts observed in EEG signals [6]. It introduces noise in the range of 15–20 Hz while removing eye blinking artifacts. Normally the WTmethod is used for artifact removal. However, this method is relatively slow and cannot be used to separate components overlapping in time-frequency space. Thus the extraction of singular spectrum contributes to the removal of artifacts from EEG signals and smoothening spectrum segmentation.

The singular spectrum (SS) method is used for applications such as EEG-based sleep analysis [7,8], artifact removal from a single channel of an EEG signal [9], brain–computer interface (BCI), and extracting brain rhythms for disease detection [10].

The SS method is a statistical tool which segments the time-space of the signal, separates noise from the signal, and smoothens the effect of the filters [11]. The flow of the algorithm is shown in Fig. 5.11 The method uses SS segmentation followed by EMD and EWT decomposition [12]. The features are extracted from SSEMD-IMF1 and SSEMD-IMF2, as shown in Figs. 5.12 and 5.13. Whereas Figs. 5.14 and 5.15 demonstrate features extracted from SSEWT-IMF1 and SSEWT-IMF2. These features are used by classifiers for seizure classification with improved accuracy.

The extracted features are observed to be more specific for all the feature values due to refinement in decomposition by SS analysis. Fig. 5.12(a) shows the entropy features extracted from SSEMD-IMF1 and SSEMD-IMF2 for all the classes. It is observed that there is distinct variation in the feature values for each of the classes. Similarly, the PSD, ZCR, and standard deviation values are separable for each class, as shown in Fig. 5.12(b)–(d). The spectral features also have different values for each class, as plotted in Fig. 5.13(a) and (b) for SS and SF. It is clear that the SSEMD-based features are more relevant and can improve the classification accuracy. Fig. 5.14(a) and (b) shows spectral entropy and ZCR features extracted for all the classes. Spectral entropy is a separable feature using SSEWT and ZCR also shows separation with the same trend for each class. Fig. 5.14(c) demonstrates PSD values for all classes; it is observed that the PSD values for the MYSZ, TNSZ, and nonepileptic classes are distinguishable from those in other classes; thus, PSD val-

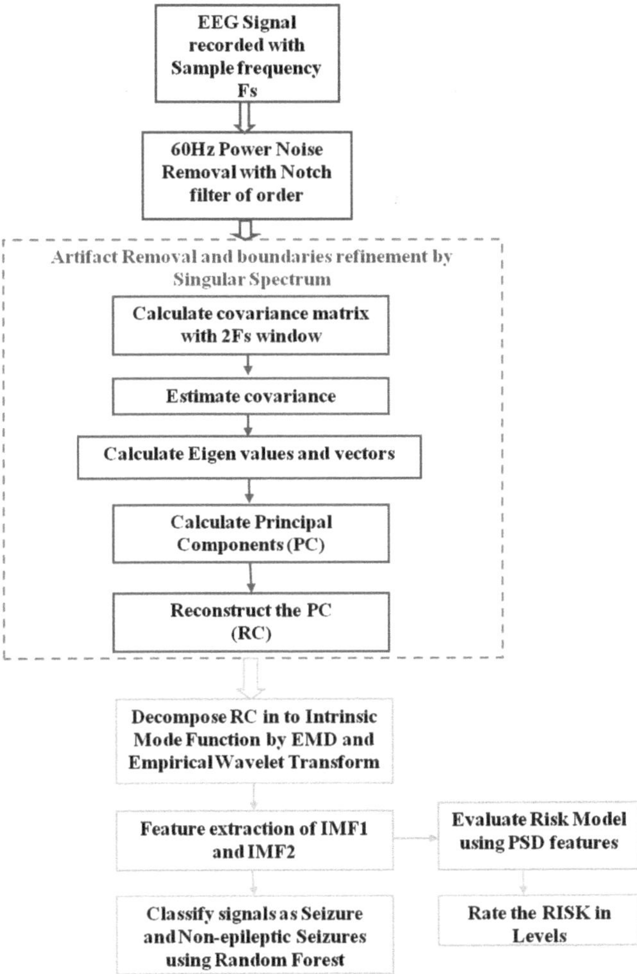

Figure 5.11 Effective method for singular spectrum EMD or EWT to improve classification.

ues are considered for the classification of epileptic and nonepileptic patients.

Fig. 5.14(d) shows standard deviation features; it is observed that most of the features are separable and thus useful for classification. Fig. 5.15(a) and (b) shows spectral features extracted for IMF1 and IMF2. The separable characteristic of feature values makes them relevant, as confirmed by ANOVA in the previous section.

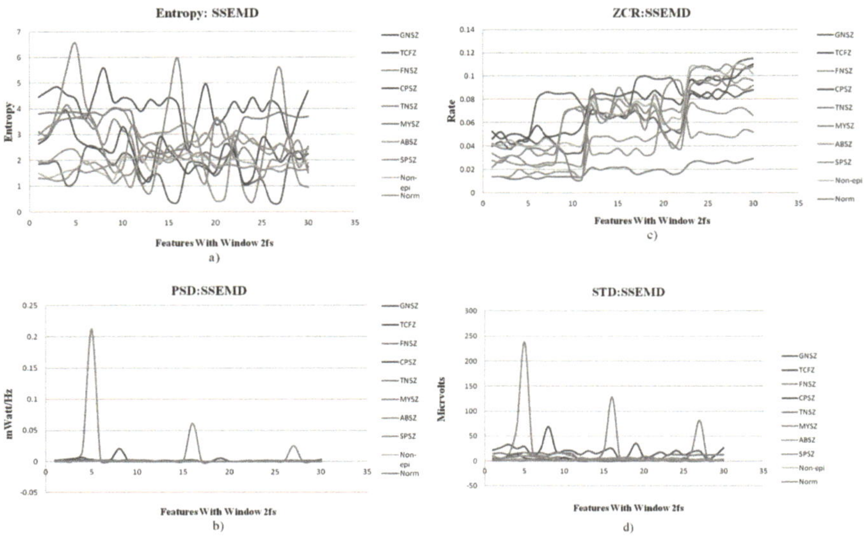

Figure 5.12 SSEMD features: entropy, PSD, ZCR, and standard deviation (STD).

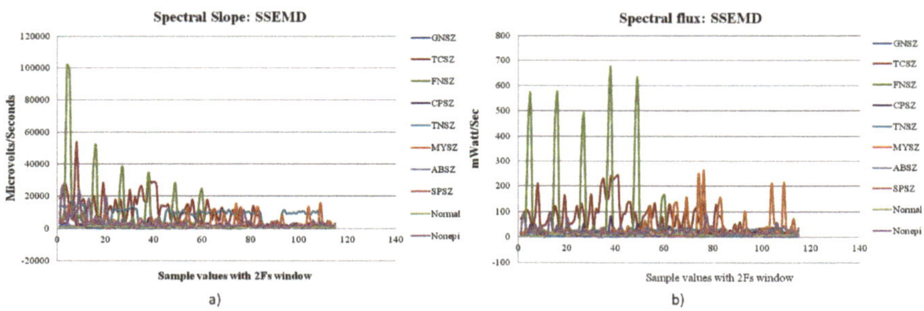

Figure 5.13 SSEMD features: spectral slope (SS) and spectral flux (SF).

5.4 Classification using machine learning classifiers

The classification algorithms are used to determine to which epileptic or nonepileptic category and subclass certain EEG signals belong based on extracted features. This step includes a strategy for determining prominent features to be selected and combining feature vectors constructed to optimize performance. The seizures are classified using a random forest tree-based classifier.

It is a supervised learning method. As discussed, 60% of the files in dataset are used for training the classification model for all the class types,

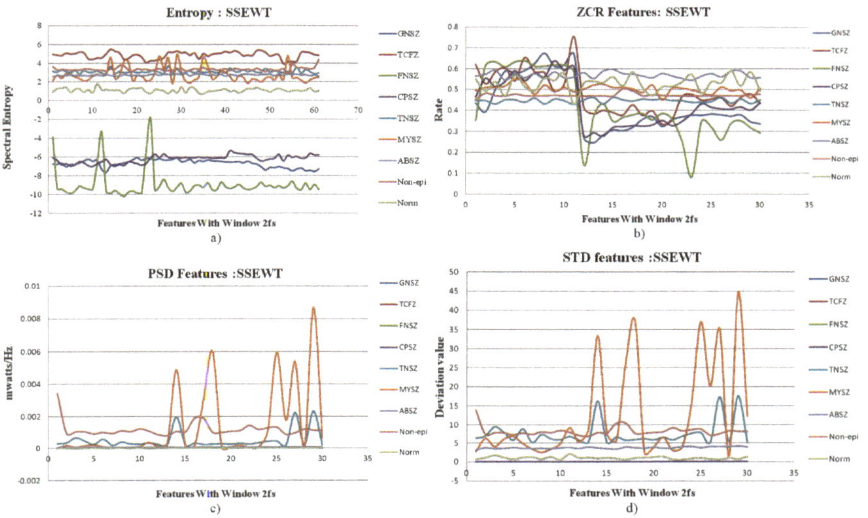

Figure 5.14 SSEWT features: entropy, PSD, ZCR, and standard deviation (STD).

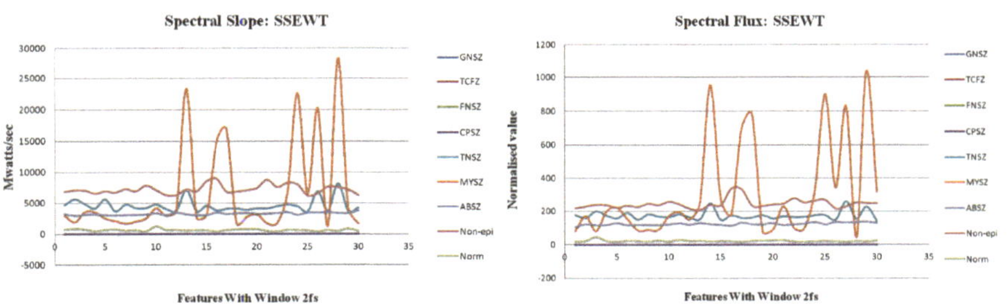

Figure 5.15 SSEWT features: spectral slope (SS) and spectral flux (SF).

as mentioned in Table 5.1, and 40% of the files are used for testing the validation method with K-fold validation and 60% cross-validation. The classifier is implemented with the well-established machine learning tool WEKA. The results are discussed in the last section.

As per the decision tree principle, the model is trained based on data, and the predictive model is built to map the tree-based decision. Given the training dataset features and specified classes, the decision tree algorithm will set rules for classification. The main limitation of the decision tree is the selection of the "best split parameter" for the prediction decision. The use of the hybrid concept of ensemble and decision tree classification, i.e., a random forest classifier, can address this limitation. Random forest is a

- F1(TFD)= f (Var, PSD, ZCR, Entropy)
- F2(WT)= f (PSD_gamma, ZCR_gamma, PSD_beta, ZCR_beta)
- F3(SSEWT)= f (PSD_gamma, ZCR_gamma, SS_gamma, SF_gamma, SE_gamma, PSD_beta, ZCR_beta, SS_beta, SF_beta, SE_beta)
- F4(SSEMD)= f (I1_En, I1_PSD, I1_STD, I1_ZCR, I2_En, I2_PSD, I2_STD, I2_ZCR)
- F5(SSEMD)=f (I1_En, I1_PSD, I1_STD, I1_ZCR, I1_SS, I1_SF I2_En, I2_PSD, I2_STD, I2_ZCR, I2_SS, I2_SF)
- **I1 and I2 are IMF1 and IMF2

Figure 5.16 Mathematical equations for feature vectors used for classification.

supervised learning algorithm. It is an ensemble of a decision tree-based method trained with the bagging method.

Random forest constructs various decision trees and combines them for accurate and stable prediction. One significant advantage of the random forest classifier is that it is useful for classification and regression problems, and hence it is preferred in current machine learning systems. It explores the most relevant features among a randomly generated subset of features, thus describing the optimized model. The score of each feature is automatically generated after training, and according to this score, the features are selected for classification. Fig. 5.16 shows the feature vector construction used for classification using EWT, SSEMD, and SSEWT after feature selection.

5.5 Results and discussion

The performance metrics analyzed here are the true positive rate (TPR), sensitivity, false positive rate, precision, and recall, and the receiver operating curve (ROC) shows the relation between sensitivity (recall) and the false positive rate.

Table 5.3 shows that the maximum classification for each seizure class is observed with EWT, except TCSZ and MYSZ. The DWT gives comparatively good accuracy for TCSZ and MYSZ.

The false positive rate is shown in Table 5.4. It is observed that EMD has a better false positive rate than WT, but WT has better classification accuracy than EMD. EWT uses a hybrid model of WT and EMD and exhibits a better classification accuracy along with a false positive rate.

Table 5.3 Classification accuracy for WT, EMD, and EWT.

Class	Wavelet transform		Empirical mode decomposition		Empirical wavelet transform	
	10-fold	70% validation	10-fold	70% validation	10-fold	70% validation
Normal	89.4	89.9	90.2	91.7	**94.3**	93.4
Nonepi	55.5	53.3	63.5	66.2	**67.6**	64.4
GNSZ	29.3	22.2	45.1	42.9	**62**	60.3
TCSZ	87.9	86.3	85.5	79.1	69.7	71.4
FNSZ	31.2	32.3	49.3	50.4	**80.2**	81.3
CPSZ	35	35.3	67	61.9	**88.7**	89.6
TNSZ	92.4	90.5	89.9	90.8	**96.4**	92.6
MYSZ	56.3	54.6	68.4	67.1	32.9	37.6
ABSZ	45	41.1	64.3	61.1	**91.4**	89.1
SPSZ	74.7	73.2	81.4	82.6	**99.6**	98.7

Table 5.4 Comparative study of false positive rates for WT, EMD, and EWT.

Class	Wavelet transform		Empirical mode decomposition		Empirical wavelet transform	
	10-fold	70%	10-fold	70%	10-fold	70%
Normal	0.026	0.025	0.032	0.04	0.009	0.008
Nonepi	0.122	0.124	0.137	0.143	0.046	0.043
GNSZ	0.01	0.014	0.018	0.019	0.004	0.003
TCSZ	0.019	0.02	0.014	0.026	0.023	0.019
FNSZ	0.052	0.059	0.049	0.054	0.013	0.009
CPSZ	0.01	0.018	0.011	0.012	0	0
TNSZ	0.04	0.035	0.036	0.035	0.002	0.004
MYSZ	0.026	0.031	0.211	0.224	0.023	0.025
ABSZ	0.053	0.051	0.008	0.01	**0.009**	0.009
SPSZ	0.087	0.092	0.019	0.02	**0**	0.001

Thus, performance analysis for EWT features states that the maximum accuracy is observed with 10-fold validation for the SPSZ class (99.6%) and the lowest accuracy is observed with 70% validation for MYSZ, whereas the classification accuracy is 67% for NES with a 5-fold method. Thus it is observed that 10-fold and 70% validation could produce significant results.

Table 5.5 Classification accuracy and false positive rate of SSEMD and SSEWT with 10-fold and 70% cross-validation.

Classification accuracy or TPR				
Class	**Singular spectrum empirical wavelet transform**		**Singular spectrum empirical mode decomposition**	
	10-fold	**70%**	**10-fold**	**70%**
Normal	96.1	96.2	94.6	94.4
Nonepi	94.2	94.9	87.6	84.5
GNSZ	95.7	94.7	93.5	85
TCSZ	99.1	99.2	99.1	99.2
FNSZ	99.6	99.5	81.1	76.9
CPSZ	85.2	93.8	95.7	92.2
TNSZ	99.4	99	96.7	98.6
MYSZ	95.5	94.1	78.9	79.7
ABSZ	98.2	97.6	81.6	75.3
SPSZ	98.7	98	99.8	99.5

Similarly, the methods are implemented for WT and EMD using random forest with 10-fold and 70% validation. The classification accuracies of these methods are shown in Table 5.3.

Table 5.5 shows a summary describing the classification accuracies achieved with SSEMD and SSEWT along with the false positive rates. SSEMD and SSEWT are primarily designed for NES detection. The SSEMD and SSEWT methods improve the classification accuracy for NES along with other classes [13]. Table 5.5 shows the classification accuracies and true positive rates for SSEMD and SSEWT.

SSEWT provides a good classification accuracy for almost all classes, with sufficiently high false positive rates, and 10-fold validation for random forest classifiers is a well-established method for seizure classification.

5.6 Risk analysis model

The mathematical model for risk analysis is derived from the literature and based on parameters such as age, partial/generalized (P/G), strength, frequency, and duration. Every signal will be analyzed for these parameters and classified as risk 1, 2, or 3, and an alert system can be further added to this research.

Table 5.6 RISK level ratings and ranges of severity and frequency.

Severity (S) = PSD strength	P = frequency of seizure occurrence (F) × total duration (T)	Risk (R = SP)
Minor (1): less than > 30% of mean	Unlikely (1): <2 min	Low (1)
Moderate (2): > between 30% and 60%	Likely (2): between 2 and 10 min	Medium (2)
Serious (3): >60%	Very likely (3): >10 min	High (3)

Table 5.7 Parameters observed for the risk model.

Patient file	Age	Seizures	PSD, strength	Seizure frequency	Duration	Level
258_s003_t000	41	FNSZ	0.00343	5	130.30	0
4671_s007_t000	27	GNSZ	0.0290	39	7.60	1
6546_s014_t000	40	GNSZ	8.376	9	71.30	3
6546_s018_t000	41	FNSZ	0.113	7	113.10	2

- Input: Fi = (age, P/G, strength, frequency, duration)
- Weight = (0.1, 0.1 × (0, 1), 0.2, 0.3, 0.3)
- Weights decided from database type and recording method
- Output: risk levels

Table 5.6 states the considered ranges of severity and frequencies of occurrence to identify the risk level associated with the received signal.

The risk analysis model is built on various parameters and shows various risk levels (Table 5.7). Longer durations of seizures and higher frequencies are associated with higher risk.

The graphical user interface (GUI) is designed using a MATLAB tool which has two options of signal input: *.txt and *.edf. The output is classification of the signal as normal or seizure together with its subcategory. The alert messages will be displayed on the GUI screen. The GUI is shown in Figs. 5.17 and 5.18.

5.7 Summary

The present work discusses the classification of EEG signals as normal and seizures, with various subclasses, using WT, EMD, EWT, SSEMD, and SSEWT. SSEWT is an efficient method for classification. The risk assessment model is designed with GUI. The GUI and algorithms can be

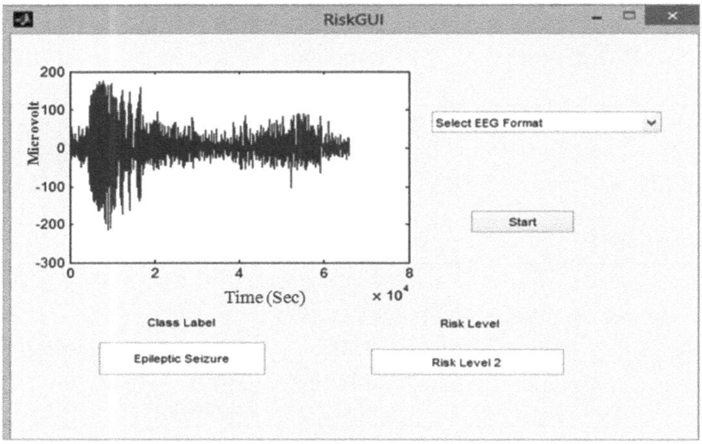

Figure 5.17 GUI showing detection of seizures and their levels.

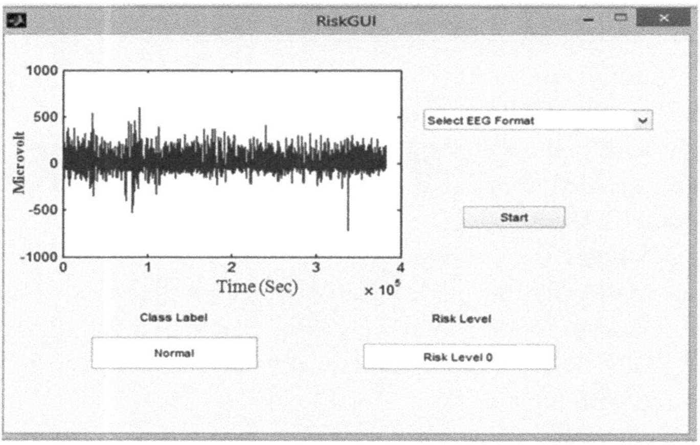

Figure 5.18 GUI showing detection of normal signals and their levels.

embedded in existing systems for classification of EEG signals into specific subclasses discussed here.

References

[1] A. Goldberger, L. Amaral, L. Glass, J. Hausdorff, P.C. Ivanov, R. Mark, et al., PhysioBank, PhysioToolkit, and PhysioNet: components of a new research resource for complex physiologic signals, Circulation 101 (23) (2000) e215–e220 [Online].

[2] M. Golmohammadi, V. Shah, S. Lopez, S. Ziyabari, S. Yang, J. Camaratta, J.

Obeid, J. Picone, The TUH EEG seizure corpus, in: American Clinical Neurophysiology Society, Phoenix, Arizona, USA, 2017, p. 1 [Online]. Available https://www.isip.piconepress.com/projects/tuh_eeg/. (Accessed 7 November 2017).
[3] Angela Zeiler, Rupert Faltermeier, Ingo Keck, Ana Tomé, Carlos Puntonet, Elmar Lang, Empirical mode decomposition – an introduction, in: Proceedings of the International Joint Conference on Neural Networks, 2010, pp. 1–8.
[4] J. Gilles, Empirical wavelet transform, IEEE Transactions on Signal Processing 61 (16) (2013) 3999–4010.
[5] A. Bhattacharyya, R.B. Pachori, A multivariate approach for patient-specific EEG seizure detection using empirical wavelet transform, IEEE Transactions on Biomedical Engineering 64 (9) (2017) 2003–2015, https://doi.org/10.1109/TBME.2017.2650259.
[6] V.K. Harpale, V.K. Bairagi, Significance of independent component analysis (ICA) for epileptic seizure detection using EEG signals, in: Proceedings of the International Conference on Data Engineering and Communication Technology (ICCUBEA2017), Pune, in Advances in Intelligent Systems and Computing, vol. 2, 2017, pp. 829–838.
[7] S. Enshaeifar, S. Kouchaki, C.C. Took, S. Sanei, Quaternion singular spectrum analysis of electroencephalogram with application in sleep analysis, IEEE Transactions on Neural Systems and Rehabilitation Engineering 24 (1) (2016) 57–67.
[8] S. Aydın, H.M. Saraoglu, S. Kara, Singular spectrum analysis of sleep EEG in insomnia, Journal of Medical Systems 35 (4) (2009) 457–461.
[9] A.K. Maddirala, R.A. Shaik, Motion artifact removal from single channel electroencephalogram signals using singular spectrum analysis, Biomedical Signal Processing and Control 30 (2016) 79–85.
[10] L. Spyrou, Y. Blokland, J. Farquhar, J. Bruhn, Singular spectrum analysis as a preprocessing filtering step for fNIRS brain computer interfaces, in: Proceeding of 22nd European Signal Processing Conference (EUSIPCO), Lisbon, 2014, pp. 46–50.
[11] H. Hu, S. Guo, R. Liu, P. Wang, An adaptive singular spectrum analysis method for extracting brain rhythms of electroencephalography, Peer Journal 5 (2017) e3474.
[12] Vinayak K. Bairagi, Varsha K. Harpale, Improved epileptic seizure detection using singular spectrum empirical mode decomposition and machine learning approach, Journal of Statistics and Management Systems (2021), https://doi.org/10.1080/09720510.2020.1862958.
[13] V. Harpale, V. Bairagi, S. Kolat, V. Bendre, EEG based computationally optimized solution for non-epileptic seizure detection, in: Proceeding of 4th International Conference on Communications and Cyber-Physical Engineering, Pune, 2021.

CHAPTER 6

Concluding remarks and future scope

6.1 Seizure detection results

As discussed earlier, the methodology adopted for research is feature extraction and feature selection using analysis of variance (ANOVA). With the application of optimized classifiers the system is implemented for good accuracy.

Time-frequency domain (TFD)-, wavelet-, and empirical mode decomposition–intrinsic mode function (EMD-IMF)-based features are classified using various classifiers such as naive Bayes, multiclass, and J48 with 10-fold and 66% split modes of classification [2]. Cross-validation is a technique to evaluate predictive models by partitioning the original sample into a training set to train the model and a test set to evaluate it.

In k-fold cross-validation, the original sample is randomly partitioned into k equal-size subsamples. The classifiers used in this work use 10-fold cross-validation for partitioning the feature set in random testing and training datasets to verify their performance. Split validation is a way to predict the fit of a model to a hypothetical testing set when an explicit testing set is not available. The split validation operator also allows training on one dataset and testing on another explicit testing dataset. The work uses 66% of the total dataset for training and the remaining data for testing.

As shown in Table 6.1, performance metrics or confusion metrics are derived for EMD and singular spectrum (SSEMD). It is observed that the best results are obtained with SSEMD-based seizure detection. The new SSEMD model shows better the seizure detection accuracy as 99.8% and reduces the false positive rate (FPR) to 0% in the classification of normal and epileptic patients as compared to existing proven methods, and it helps improving early detection of epilepsy. The new electroencephalography–genetic algorithm (EEG-GA) architecture can easily be used in a real-time system as an ASIC solution. The system can achieve an accuracy of about 96% for seizure detection. Dimensionality reduction is observed by about 40% of total dimensions with ANOVA, and a hybrid model of ANOVA+GA gives a 63% reduction of feature space [3,4]. The design

Brain Seizure Detection and Classification Using EEG Signals
https://doi.org/10.1016/B978-0-32-391120-7.00011-6

Table 6.1 Performance metrics observed using wavelet transform.

Method classifiers	TPR	FPR	P	F1	ROC
10-fold naive Bayes	0.84	0.02	0.878	0.836	0.872
10-fold multiclass classifier	0.93	0.15	0.911	0.92	0.918
10-fold J48	0.95	0.09	0.812	0.95	0.946
Naive Bayes (66% per split)	0.917	0.19	0.791	0.845	0.953
Multiclass classifier (66% per split)	0.932	0.22	0.904	0.913	0.927
J48 (66% per split)	0.995	0.09	0.893	0.935	0.969

could provide low power (37 mW) and less complexity or improved performance platform-specific reconfigurable solutions, contributing to the design of real-time seizure detection systems.

6.2 Seizure prediction results

Quantitative EEG analysis is a significant tool to extract timing parameters and a frequency component. A suitable method of analyzing highly dynamic, nonlinear, and nonstationary EEG signals entails the use of flexible and nonsubjective measures. The system implemented here is trained for various TFD features, and it is observed that EEG signals of various patients have characteristics of increased amplitude and frequency variation [6]. The signals were analyzed based on features and with a fuzzy classifier for better performance, and the presented method could achieve a classification accuracy of 96.48%, a TPR of 96.52%, and an FPR of 0.352 for seizure detection and an accuracy of 96.02% for identifying the preseizure state [1].

As shown in Table 6.2, the preseizures are detected for all 22 patients, and it is observed that the maximal preseizure time detected for the patient 6 database is 110 seconds before the arrival of seizures. The preseizure state is detected at 13 to 110 seconds before the actual seizure onset, which can further improve by concentrating on related features. The fuzzy logic classifier model is used to classify transitions between brain states. The preseizure time observed in each case is different, so on average seizures are predicted approximately 30 sec before occurrence. The accuracy is improved with the implemented fuzzy inference system.

The performance of the system for preseizure detection can be improved by extracting relevant features. Due to variation in seizure charac-

Table 6.2 Prediction observed for preseizure time for patient 3 and patient 6.

Seizure signal	Patientfile	Seizure time (sec)	Results preseizure time (sec)	Early detection time (sec)
1	chb03_01.edf	362	348	14
2	chb03_02.edf	731	714	17
3	chb03_03.edf	432	398	34
4	chb03_04.edf	2162	2102	60
5	chb03_34.edf	1982	1937	45
6	chb03_35.edf	2592	2508	90
7	chb03_36.edf	1725	1712	40
8	chb06_01.edf	1724	1698	26
9	chb06_01.edf	7461	7405	56
10	chb06_01.edf	13525	13489	36
11	chb06_04.edf	327	267	60
12	chb06_04.edf	6211	6114	97
13	chb06_09.edf	12500	12390	110
14	chb06_10.edf	10833	10762	71
15	chb06_13.edf	506	441	65
16	chb06_18.edf	7799	7695	104
17	chb06_24.edf	9387	9293	94

teristic training, system design is a challenging task. Further, the system performances can be enhanced by selecting very precise features, an advanced and most suitable classifier, and selecting a high-performance nonspecific algorithm. Prediction of seizures well before seizure onset still requires further research.

6.3 Seizure classification results

The features are extracted using existing methodology such as TFD, WT, EMD, and EWT and features are extracted from proposed methods SSEMD and SSEWT [2]. The feature selection is performed with one-way ANOVA (wrapper method of feature selection). The decision tree-based algorithm random forest classifier is used for classification in classes discussed in the seizure prediction. But the main objective is to classify seizures as epileptic seizures and nonepileptic seizures.

Table 6.3 Summary of classification accuracy using EWT.

Class	Random forest	Accuracy
Normal	10-fold	94.3%
Nonepi	5-fold	67%
GNSZ	5-fold	62.6%
TCSZ	70% validation	71.4%
FNSZ	70% validation	81.3%
CPSZ	70% validation	89.6%
TNSZ	10-fold	96.4%
MYSZ	70% validation	37.6%
ABSZ	10-fold	91.4%
SPSZ	10-fold	99.6%

Decision trees are classic supervised learning algorithms, easy to understand and easy to use. The main concept behind decision tree learning is the following: starting from the training data, we will build a predictive model which is mapped to a tree structure. The goal is to achieve perfect classification with a minimal number of decisions, although this is not always possible due to noise or inconsistencies in data.

For seizure classification, the classification accuracy varies as per class features extracted using empirical wavelet transform (EWT); for example, it is 99.6% and 37.6% for SPSZ and MYSZ, respectively. The EWT method can classify normal and nonepileptic seizures, with just 66.7% of accuracy. Hence a new method is proposed, and SSEMD and SSEWT are used to improve classification accuracy. It is observed that SSEWT provides good classification accuracy and FPR as compared to all other experimented methods to detect nonepileptic seizures. The EWT gives an accuracy of 87.6% with an FPR of 0.007. The work is further extended by defining a risk model for detected seizures and generating an alarm for severity. Table 6.3 shows a summary of experimentation results using EWT; an accuracy of 67% was observed for nonepileptic seizure classification. The work was extended with the new methods SSEMD and SSEWT, which provide improved accuracy for nonepileptic seizure classification. Thus the classification accuracy is improved with SSEMD to 87.6% with 10-fold and 84.5% with 66% cross-validation and with SSEWT to 94.2% with 10-fold and 94.9% with 66% cross-validation. Thus the method achieves better performance.

Table 6.4 Complexity calculation for existing and proposed methods.

Method	Feature space	Time complexity	Computational complexity (Big (O))
	Existing methods		
DWT	13×11 windowed samples ×5 freq. bands×100 files = 71,500 instances	O(log(m)log(n)) m=5; n=11×100	J48: $O(V^2)$ V=13
EMD	13×11 windowed samples ×5 frequency bands ×100 files = 71,500 instances	Ns(41m×n) Ns=23; m=5, n=11×100	$O(V^2)$ V=13
	Presented work		
SSEMD	8×11 windowed samples ×2 frequency bands ×100 files = 17,600 instances	Ns(41m×n) Ns=23; m=2, n=11×100	O(V×n (log(n))) V=8
SSEWT	6×11 windowed samples ×2 frequency bands ×100 files = 13,200 instances	O(log(m)log(n)) m=2; n=11×100	O(V×n(log(n))) V=6

6.4 Summarized observations

The research conducted specifically contributes to seizure detection, seizure prediction, and seizure classification by presenting new methods of feature extraction. The SSEMD and SSEWT feature extraction methods have improved decomposition bandwidth. The overall time complexity and computational complexity of SSEMD and SSEWT are calculated and compared with those of the existing methods DWT and EMD. Table 6.4 shows the actual feature space observed with DWT and EMD as well as the reduced feature space due to application of feature selection (one-way ANOVA). As overall complexity depends on decomposition levels (m) and feature attributes (V), it is reduced as compared to existing methods.

The overall results obtained in the present work for seizure detection, prediction, and classification are stated in Table 6.5. The major objectives of the work are seizure prediction and classification of epileptic and nonepileptic seizures to avoid unnecessary application of antiepileptic drugs. The work is compared with published work of other researchers reviewed in Paul et al. [7].

The research conducted achieved 99.8% accuracy with an FPR of 0.001 for seizure detection, 94.75% seizure prediction accuracy, and 94.9% classification accuracy, classifying epileptic and nonepileptic seizures.

Table 6.5 Performance comparison of presented research work.

Database (classes)	Review by Paul et al. [7]			Presented work		
	Method + classifier	**Acc. (%)**	**FPR**	**Method + classifier**	**Acc. (%)**	**FPR**
CHB-MIT (02)	Time+ GA	98.58	0.015	PATW + Fuzzy [1]	95.94	0.052
CHB-MIT (03)	Wavelet+ FIS	93.57	0.05	PATW+ Fuzzy [1]	94.75	0.052
Andrezejak 2001 (02)	EMD+ linear Bayes	98	–	SSEMD+ NB, J48, MC** [2]	99.8	0.001
TUH (10)	–	–	–	SSEWT+ RF	94.9	0.003

MC**: Multiclass classifier.

6.5 Conclusions

EEG is a noninvasive method used to record brain activity; it is simple and inexpensive, and has a good temporal resolution. The disadvantages of EEG are that the signal-to-noise ratio is low, thus demanding extensive data analysis, poor spatial resolution, and the requirement for a large subject-specific database. Frequency analysis of EEG signals is conducted to extract quantitative parameters and compare between the power at different frequency bands and their topological distribution. Time-frequency analysis is used in various applications such as automatic seizure detection. Thus EEG is the preferable diagnosis tool for analyzing and detecting various brain disorders.

Very favorable methods of EEG feature extraction are based on the fast Fourier transform, short-time Fourier transform, and discrete wavelet transform (DWT), out of which wavelet-based analysis is most commonly used. Recent trends are derived from a combination of the previously developed analysis methods to overcome their limitations and improve their performance, such as sensitivity, specificity, and classification accuracy.

Independent component analysis (ICA) is a well-established artifact removal method. Combination of ICA and other methods such as ESD, entropy, and kurtosis improves the signal-to-artifact ratio and ensures corrected signal without artifacts. ICA can also be used for epileptic seizure detection, but accuracy improvement still is a challenge in this method. It was reported that system-on-chip (SoC) and FPGA implementation of the ICA method for seizure detection achieves the greatest accuracy [5]. A real-time seizure prediction and detection system could lead to better treatment and safety for epileptic and nonepileptic patients. EEG is a noninvasive tool, which can provide high temporal resolution information assessing dynamic brain activity. In this work we detect seizures and classify them as epileptic or nonepileptic. Moreover, an automatic prediction model is designed, which is required for correct detection, continuous monitoring, and reducing operational cost. This work was conducted to achieve improved performance parameters such as significant (more than 90%) detection and classification accuracy, sensitivity, and specificity, and high prediction rate. Detected seizures will further be analyzed for severity. Hardware implementation of the proposed algorithm will facilitate the development of portable devices.

The EEG signals are analyzed with time, frequency, and time-frequency methods such as wavelet transform and EMD. The proposed work used time and frequency information and various proven methods for seizure detection for experimentation such as DWT, EMD, and SSEMD for good accuracy and improved FPR. The average feature dimensionality reduction observed is about 71% of total dimensions of features.

There are various well-established methods for artifact removal. The present work uses ICA for EEG artifact removal. Combination of ICA and other methods such as entropy and kurtosis improves the signal-to-artifact ratio and ensures corrected signal without artifacts. The proposed EEG-GA architecture facilitates to use it in a real-time system as an ASIC solution. The system can achieve an accuracy of about 96% for seizure detection. Dimensionality reduction of about 40% of total dimensions is observed with ANOVA, and a hybrid model of ANOVA+GA gives a 63% reduction of feature space. The ASIC-based design implemented on FPGA could provide low power (37 mW) and less complexity or improved performance platform-specific reconfigurable solutions, contributing to the design of real-time seizure detection systems.

A new method, SSEMD, is proposed for seizure detection, and the accuracy observed is 99.8%, with an FPR of 0.001 using SSEMD as a feature extraction method with ANOVA-based feature selection and a multiclass machine learning-based classifier. The risk model is defined and tested with power spectral density feature of IMFs and other parameters.

For seizure prediction the signals were analyzed with TFD and pattern adapted wavelet features and classified with a fuzzy classifier model to obtain better performance, and the presented method achieved a classification accuracy of 95.94% and an FPR of 0.052 for seizure detection and an accuracy of 94.75% for preseizure state identification. The preseizure state is identified at 13 to 110 seconds before the actual seizure onset, which can be further improved by optimizing related preseizure features.

For seizure classification, the classification accuracy varies as per class features extracted using EWT; for example, it is 99.6% and 37.6% for SPSZ and MYSZ, respectively. The contribution of this work lies in the identification of nonepileptic patients using SSEMD and SSEWT [8]. The work achieved a maximum accuracy of 99.8% for SPSZ with an FPR of 0.001 and 94.9% for nonepileptic patients with an FPR of 0.003.

6.6 Future scope

This work was implemented for providing a solution for early seizure detection in epileptic and nonepileptic patients. The work includes EEG signal analysis for artifact removal, seizure detection, seizure prediction, and seizure classification. Thus the scope of this work is limited to proposing the new algorithm SSEWT, its performance analysis, and the development of a risk model for severity analysis of seizures. This work can be further extended to:

1. design an accurate wireless seizure monitoring and alert system to help doctors and epileptic patients in rural areas,
2. develop a prediction model for early diagnosis of epileptic seizures,
3. classify EEG signals according to disease symptoms,
4. develop an algorithm for seizure classification and to include it in existing EEG monitoring systems, and
5. contribute to society by defining the large-scale probability of brain disorders like autism, as autistic patients may suffer from epilepsy.

References

[1] Varsha Harpale, Vinayak Bairagi, An adaptive method for feature selection and extraction for classification of epileptic EEG signal in significant states, Journal of King Saud University: Computer and Information Sciences (ISSN 1319-1578) (2018), https://doi.org/10.1016/j.jksuci.2018.04.014.

[2] Vinayak K. Bairagi, Varsha K. Harpale, Improved epileptic seizure detection using singular spectrum empirical mode decomposition and machine learning approach, Journal of Statistics & Management Systems (2021), https://doi.org/10.1080/09720510.2020.1862958.

[3] Varsha Harpale, Vinayak Bairagi, FPGA based architecture implementation for epileptic seizure detection using one way ANOVA and genetic algorithm, Biomedical and Pharmacology Journal 12 (3) (2019) 1543–1553.

[4] V.K. Harpale, V.K. Bairagi, S.S. Pujari, Real time SoC architectures for analysis of EEG, in: Proceeding of 3RD IEEE, International Conference on Computing, Communication, Control and Automation, 2017, pp. 1–5.

[5] V.K. Harpale, V.K. Bairagi, Significance of independent component analysis (ICA) for epileptic seizure detection using EEG signals, in: Proceedings of the International Conference on Data Engineering and Communication Technology (ICCUBEA2017), Pune, in Advances in Intelligent Systems and Computing, vol. 2, 2017, pp. 829–838.
[6] V.K. Harpale, V.K. Bairagi, Time and frequency domain analysis of EEG signals for seizure detection, in: Proceeding of the IEEE International Conference on Microelectronics, Communication & Computing (MicroCom 2016), NIT, Durgapur, 2016.
[7] Y. Paul, Various epileptic seizure detection techniques using biomedical signals: a review, Brain Informatics 5 (2) (2018).
[8] V. Harpale, V. Bairagi, S. Kolat, V. Bendre, EEG based computationally optimized solution for non-epileptic seizure detection, in: Proceeding of 4th International Conference on Communications and Cyber-Physical Engineering, Pune, 2021.

Index

Kolibri 360 GmbH
Ferdinand-Jühlke-Straße 7
99095 Erfurt, Deutschland
produktsicherheit@kolibri360.de